I0755663

# BROOKLYN WATER WORKS

## PUMPING ENGINE No. 1.

| | |
|---|---|
| Diameter of Cylinder | 90" |
| Length of Stroke of Piston | 10' |
| Number of Pumps | 2 |
| Dia. of working barrel | 36" |
| Do. " Auxilliary do | 54" |
| Pump Stroke | 10 |
| Do. Valves of the Double beat kind | – |
| Double acting Air Pump | – |
| Diameter of Do. | 3' |
| Length of Stroke | 5' |
| Height of Centre of Beam above level of Floor | 26'3" |
| Length of Do. between end centres | 30' |
| Depth of Do. in middle | 7'2" |
| Averrage thickness of Web | 0'6" |
| Diameter of Air Chamber | 6'8½" |
| Height of Air Chamber | 25'4" |
| Do. do. do. above Floor | 13'10" |
| Total weight of Engine, Boilers & Appurtenances | 440 Tons |

WOODRUFF & BEACH, IRON WORKS
Hartford Ct. Jany. 27th 1860.

Patented Novr. 15th 1859 by Wm. WRIGHT Hartford Ct.
Drawn by F. Budden.

D. Van Nostrand, Publisher 192 Broadway, N.Y.
J. Bien Lith. N.Y.

THE

# BROOKLYN WATER WORKS

AND

# SEWERS.

A

## DESCRIPTIVE MEMOIR.

---

PREPARED AND PRINTED BY ORDER OF

THE BOARD OF WATER COMMISSIONERS.

ILLUSTRATED BY FIFTY-NINE LITHOGRAPHIC PLATES.

NEW YORK:

D. VAN NOSTRAND, 192 BROADWAY.

1867.

# CONTENTS.



## APPENDIX.

# LIST OF PLATES.

# INTRODUCTION.

The necessity for an abundant supply of pure water to large cities has never been questioned. Wherever a dense population is gathered together on a limited area the slender amount derived from wells or cisterns is liable to be rendered impure, while the labor of raising it deters many from using water freely for sanitary or household purposes. A plentiful and unfailing public supply furnishes also the means for cleansing sewers, and the prompt quenching of fires. Manufactories, public baths, and shipping require large quantities of pure water, at a moderate cost. The certainty of having it thus furnished, attracts and fixes much business which would otherwise be scattered over the country.

All these requirements can only be met by a supply delivered with such a head that the water will rise freely to the highest portions of buildings, even after the loss of much of its pressure while passing through the mains and branches.

Should a pure stream or lake at a sufficient elevation, be found near a city, the water can be delivered by a gravitating supply, as in the case of the Croton water in the city of New York. If, however, the level of the source will not permit this, recourse must be had to artificial means, in order to raise the water into the distributing reservoirs. In either case supply ponds must be prepared, in order to insure a constant level of delivery into the conduits or mains.

When a gravitating supply is not possible or advisable, the structures for conducting it may be made simpler, for the conduits may be placed at much lower levels, and costly aqueducts, such as the one over the Harlem river near New York, may be avoided. A proper place for the engine which is to raise the water, is to be selected as near as possible to the distributing reservoir, in order to save loss of power by friction in the rising mains. If steam is used, the reservoir may be smaller than in the case of an intermittent engine driven by wind or water.

Among the ancients, the Romans more especially sought to obtain an abundant supply of water for their capital, and have constructed the largest known aqueducts until modern times. Nineteen aqueducts, with a gravitating supply at a moderate head, brought to the imperial city as many streams, which supplied their fountains and baths. These works, few in number previous to the time of Cæsar, were constructed by citizens anxious to win popularity, and some of them are in use at the present time. Another of them has been recently reported as nearly perfect, and capable of being repaired at a small cost. These aqueducts crossed the Campagna, or level lands around Rome, supported on high arches. The water was carried in open channels, and as few pipes as possible were used for the distribution. It was a natural prejudice in favor of conduits over leaden pipes, which were expensive and injurious to the water, that led to the construction of these gravitating supplies. The laws of hydrostatics were well understood at the time, but the manufacture of pipes of any material but lead, or perhaps copper, was not understood. These leaden pipes with bronze faucets are found under the ruins of Pompeii, and one piece of such a pipe, about eight inches in diameter, is preserved in the Museum at Naples, with water contained in the closed faucet, which has been there over eighteen hundred years.

Many supplies are furnished to the older European cities through leaden pipes, sometimes bringing the water from a great distance, and leading it to the public fountains, whence it is carried in pails to the consumers. The loss of head, owing to the small size of the pipe and imperfect construction, is very great, and the use of lead is objectionable, though if the water is not too soft it may not become much deteriorated in them.

The invention of the steam-engine, and of the method of making cast-iron pipes, has much reduced the cost of water-works. A pumping supply, delivered by eight different companies, furnishes London, with its three millions of inhabitants, with a supply about double of that of the Croton. Paris has one gravitating and several pumping supplies, the aggregate being a very meagre amount for so large a population.

A perfect pipe distribution involves a great amount of engineering skill and mechanical ingenuity. The reservoir may well be compared to the heart, and the pipe distribution to the arteries, of a living being. The valves and capillary vessels of the arteries are represented by the stop-cocks and service-pipes; while, to carry out the simile, the sewers may be likened to the veins, which carry off the impure blood after it has performed its vital functions.

When a city has determined to procure for itself an artificial supply of water, the various sources for such a supply have to be carefully examined. An examination of the quality, capacity, and permanence of the sources, and of their relative distance, together with a study of their levels and of the line of conduits to the engines or reservoirs, must be carefully made. Then the proper position for one or more reservoirs must be selected, with reference to height and convenience to the influent supply, and the effluent delivery. The track of the principal distributing mains must be laid out, so as to avoid abrupt bends, or too many and too sudden changes in them. The details of the constructions, and the precise position or size of many of them, will require more detailed surveys, calculations and designs, in case the preliminary studies have been accepted as proper to the end in view.

In the case of the Brooklyn water supply, the question was limited to a choice between a supply derived from large wells or from the small streams which water the southerly slopes of Long Island, the nearest of which is over ten miles from the heart of the city. This question was discussed for many years before the final preference was given to the supply from the pure and never-failing island streams.

Although there can be no doubt at the present time, that a city of nearly four hundred thousand souls could not derive an adequate supply from wells, there existed a very reasonable prejudice in favor of such a source of supply when the population was not one-tenth as large. Even now, where the pipe distribution has not been extended, many depend upon cisterns or the public wells for their household supply, where the water which they furnish is pure. At any point on the southerly slope of the island, and at most places on the opposite slope where strata of clay do not interfere with the free percolation of the rain-fall to the main spring, all that is necessary in order to procure a sure though slender water supply, is to sink a well to this main spring or permanent water stratum. At a depth determined by the difference between the surface level and the level of this water-bearing gravel bed, which slopes very gently upwards from the sea, the supply is reached. Near the sea the wells may be only ten feet deep, while on the higher slopes of the central ridge they have been sunk to a depth of one hundred and twenty-five feet. An allowance of about two feet to each mile of distance from the shore determines the point at which the level of fresh water in these wells will stand. They never dry up, and furnish a pure though rather hard water, pleasant to drink, and often fit for washing. No drought ever affects them in the slightest degree, and no copious rain-fall disturbs them unless a clay bed should change the course of the infiltrating water; a case which rarely occurs. Even when a steam-engine is used to pump from them, they furnish a supply proportioned to the rapidity with which the water can pass through the gravel in which it lies.

It was, therefore, from such a source that it was at first proposed to derive a general supply. Perhaps a city of thirty thousand souls could thus be amply furnished with all the water it could use, but with a rapidly increasing population it would not be safe to rely upon the numerous wells which would be required, and which would demand a separate engine at almost each well, and perhaps an equal number of stand-pipes or reservoirs. The question as to whether the supply from these wells could be depended on in case the level of the main spring was lowered by the excessive drain upon its capacity, was discussed in an able report made by the late John S. Stoddard to the Water Committee of the Common Council in 1854. His conclusions were unfavorable to the well system, and in favor of a supply derived from the streams which drain the main spring. After this the attention of the authorities seems to have been turned exclusively to this latter source of supply.

We shall briefly review the reports made to the City Council by various successive Water Committees, each one of which labored earnestly to procure the desired information in as clear and practical a form as was possible with the sometimes limited means at its disposal. To the gradually accumulated data and explorations procured and made for these committees by engineers of experience, we owe the inception of the present works. It required years to convince even many of our most practical fellow-citizens that an abundant and the only certain supply could be brought at a moderate cost, and without any natural difficulties to oppose it, from the clear and unfailing streams on the island.

The first movement of the kind was made in 1834, the year in which the village was incorporated as a city, and when it contained a population of but twenty-three thousand souls. The committee, Messrs. Gabriel Furman and James Walters, made a report, dated March 24, 1834, which is to be found on the village records. They recommended that wells should be sunk at the base of the hill on which Fort Green (now Washington Park) was situated, and that the water should be thence raised by steam-pumps to a reservoir on the hill. Their estimate for the works, including eleven miles of ten and four-inch pipe, was one hundred thousand dollars, with an annual expense of ten thousand dollars. This report was not acted on.

Although the subject was often agitated, no formal action was taken towards procuring a public water supply until 1847, when the population had become three times as large as in 1834. Messrs. D. A. Bokee, John Stansbury, and J. W. Cochran were in that year appointed a special committee for the above-named purpose, and submitted, December 20, 1847, the opinion of Major D. B. Douglass as their report. This distinguished engineer, who first projected the Croton Works for New York, after considering the various sources of a supply, recommended one drawn from wells to be sunk south of the hills, the water to be raised by steam-power to a reservoir commanding a head of thirty to forty feet above the highest houses on the heights along the river. No surveys or estimates were made, and the subject was not again examined into until 1849.

Mayor Copeland, in his address to the Common Council, made May 7, 1849, alludes to a water supply, and favors the well system. A Water Committee, consisting of Messrs. George B. Fisk, Arthur W. Benson, George Hall, William McDonald, and J. W. Cochran, made a short report, dated January 8, 1849, recommending the well system again, and submitted a chemical analysis, made by Dr. John Torrey, of the waters from wells and streams on the island. He finds $18\frac{1}{3}$ grains of solid matter in a gallon of well water from Mount Prospect, and $10\frac{8}{10}$ grains in the water from Jamaica creek; the Croton, according to Dr. Chilton, containing 4.16 grains. The Committee estimate the cost of the works at eight hundred and thirty thousand dollars. They express obligations to Messrs. William Burden and John Gracen, among others, for plans, suggestions, and information.

It is proper here to state, that Mr. Wm. Burden had before this prepared plans contemplating a supply to be derived from Jamaica creek, the first constant stream on the island east of the city, and which is now in use for that purpose. He proposed to place an engine and a stand-pipe at each stream, and to send the water in iron pipes to a pump-well at Flatbush, whence another engine was to raise it to a distributing reservoir. His plan is noticed by Mr. McAlpine, in the Water Report of 1852, page 31. To Mr. Burden the credit is due of first pointing out the true and only unfailing source of a water supply, and it is to be regretted that the Water Committee had not given more consideration to his proposals.

In 1851, another Water Committee, consisting of Messrs. Charles R. Marvin, I. H. Smith, Edward Pell, Henry R. Kent, and E. B. Litchfield, submitted the most detailed report hitherto prepared, under date of December 20, 1851. An appropriation had been made to defray the expenses of preliminary surveys, which enabled this committee, headed by its active and able chairman, to take hold of the matter in earnest. They consulted the distinguished engineers, William J. McAlpine and John B. Jervis, who had respectively had charge of the construction of the Albany and New York Water Works. Mr. McAlpine, after examining the various proposed means of supply, considers that derived from the island streams as the only one of permanent value. He had gaugings made of the streams, commencing October 11, when they were unprecedentedly low, with the following results:

| Name of Stream. | Miles from Fulton Ferry. | Elevation of Water above High Tide. | Estimated Flow in Galls. each 24 hours. |
|---|---|---|---|
| Baiseley's (Jamaica creek) | 13¾ | 8 feet. | 5 millions. |
| P. Nostrand's | 15¼ | 13 | 1½ |
| Springfield creek | 15¾ | 10 | ¼ |
| Simonson's | 16 | 15 | 4 |
| Shaw's | 17¼ | 9 | ½ |
| P. Cornell's | 18½ | 12 | 2 |
| Pine's | 20¼ | 9 | 2 |
| L. Cornell's | 21 | 10 | 8 |
| Seeley's | 23 | 6 | ¼ |
| Baldwin's | 23¼ | 1 | ¼ |
| Millburne's | 23¾ | 7 | 2 |
| J. Smart's | 25½ | 18 | 5 |

He describes what the works would require to be, and acknowledges his obligations to Messrs. John S. Stoddard, L. S. Nash, and Ed. H. Tracey, for engineering assistance in making the exploring survey.

He proposed to place dams on the first four of the streams above noted, in order to insure a supply of ten millions of gallons a day, and by means of a conduit collect their waters in Baiseley's pond (now Jamaica reservoir), at a level of eleven feet six inches above tide. Thence a conduit was to carry the water nine miles, in cuttings varying from four to twenty feet deep, to a pump-well, whence it was to be raised to a reservoir of oval form, containing sixty millions of gallons, on Mount Prospect, one hundred and ninety-one feet above mean tide. Cornish pumping engines are proposed, one only to be erected at first, or two small ones, with united power sufficient to raise ten millions of gallons two hundred feet in twelve hours. The rising mains were to be thirty inches in diameter, and from six to ten thousand feet long. He estimates that seventy-five miles of pipe would be required for immediate use. The cost of the whole work he estimates at three and a half millions of dollars, and that works supplying six millions of gallons only, with capacities for fifteen millions, could be constructed for two and a half millions.

These plans and specifications show great skill and discretion on the part of Mr. McAlpine, and are more fully developed in the report of 1852. Had they been carried out however, the city would have been called upon to enlarge them almost immediately, so rapid has been its growth.

Mr. John B. Jervis made a short report to the Committee, not based on surveys, but recommending a supply from the streams, with perhaps recourse to wells also.

Conklin Brush, Esq., who had been Mayor of the city in 1851, and who took a warm interest in the project, in his address to the Common Council, January 5, 1852, commends the subject to its most careful consideration. The committee, at the close of its report recommends that the plan of Mr. McAlpine should be submitted to a vote of the citizens, and that if the result should be favorable, a law should be procured from the Legislature authorizing the construction of the works.

An election was directed to be held on the 27th of January, 1852, but on the 19th of that month, the Special Committee on Water for that year asked for authority and time to procure more detailed plans and estimates, and recommended that the resolution for a popular vote on the subject be rescinded ; all this was done, and the surveys were prosecuted with more attention to detail.

The Water Committee of 1852—Messrs. Chas. R. Marvin, Abraham B. Baylis, Montgomery Queen, George W. Stillwell, and Lemuel B. Hawxhurst—received from Mr. McAlpine a full report with plans and estimates which were dated April 15, 1852, but which were not published until the close of the year. In this he reviews the whole of the hitherto proposed supplies, viz., from the Croton or Bronx rivers of Westchester county, from lakes on Long Island, from wells in the vicinity of the city, or from running streams. He then describes the plan of supply from these last, which he considers the only proper one for the purpose, and gives numerous estimates of cost, founded on a supply of from five to thirty millions of gallons daily, with the means of increasing the capacity of the works if needed. The estimates vary from two millions six hundred thousand to seven millions eight hundred thousand dollars. His estimate of thirty gallons daily to each inhabitant, and his estimate of the increase of population, are subjoined :

| | | | | |
|---|---|---|---|---|
| 1855 | for 145,275 souls | 4,357,250 | gallons | daily. |
| 1860 | for 217,913 souls | 6,537,390 | " | " |
| 1865 | for 326,869 souls | 9,805,970 | " | " |
| 1870 | for 490,303 souls | 14,709,090 | " | " |
| 1875 | for 735,454 souls | 22,063,620 | " | " |

The average daily consumption for 1865, is given in the report of the Water Department as 9,643,150, a remarkable coincidence with this estimate; but the census returns for 1865 give a population of 296,112 only, though there is reason to believe that this is much too low a figure, and that the city contained nearly 320,000 souls at the close of that year.

It is impossible here to do full justice to the excellent report of Mr. McAlpine by partial extracts from it only. The plans and drawings which accompany the report were not published, but are preserved at the office of the Water Commissioners.

The Standing Committee on Water for 1853 was as follows: Messrs. Chas. R. Marvin, A. B. Baylis, John A. Dayton, Lemuel B. Hawxhurst and John Rice.

In his address of January 3, 1853, the Mayor, Ed. A. Lambert, Esq., urges the attention of the Common Council towards the subject of a water supply, and strongly favors the plan submitted by the Committee of 1852.

On the evening of January 21, 1853, an interesting debate took place in the Common Council as to whether the city should oppose the application to the Legislature of the Williamsburgh Water Company, mentioned further on, for an increase of capital and change of name. Alderman Marvin offered resolutions to this purpose and for the passage of the law prepared by the Water Committee, approved by the Council, and then pending in the Legislature. He reviewed the whole subject, and ably opposed the grant to a private company of the privilege of supplying water to this city.

Without entering into further details on matters which were then very warmly debated by the friends of both projects, it will be sufficient here to state that both acts were passed, rather late in the session. By order of the Common Council, the Mayor had sent to the Assembly an earnest remonstrance against the passage of the bill to increase the stock and privileges of the private company, but without effect.

The bill prepared by the city authorities, and which was passed June 3, 1853, authorized the Common Council to determine in what manner the city should be supplied with water, and directed the holding of a special election in order to submit the plans last mentioned, to the vote of the citizens.

A company, under the name of the Williamsburgh Water Company, had obtained a charter from the Legislature, by a law passed April 16, 1852, which authorized it to construct works, costing not over five hundred thousand dollars, for the supply of Williamsburgh alone. This company, acting on the ascertained fact that the only proper source of a supply was from the streams on the south side of the island, proceeded to purchase several of these, thus anticipating the action of the authorities of the city of Brooklyn. It obtained an amended act, which was passed June 8, 1853, changing its name to the Long Island Water Works Company, and extending its privileges so as to make its capital three millions of dollars, and authorizing it to supply Brooklyn also with water. This company employed Gen. Ward B. Burnett as engineer; received a preliminary report from him in September, 1852, and a fuller, but still a general one, embracing estimates and plans for a larger supply, in January, 1853. His report recommends the construction of a receiving reservoir at the spot where the present one has been placed, and a conduit of increasing size from the farthest streams to the pump-well.

The election, ordered by the law of June 3d, was held on the 11th of July ensuing. 7,693 votes were cast out of a population of about 116,000, containing probably 17,000 voters. Of these 2,639 were favorable to the proposed plan, and 5,054 were against it. This stopped all further progress towards procuring a water supply by public works for the time.

The plan laid before the electors was as follows: On the island streams dams were to be constructed, and the water brought in an open canal and conduit to the pump-well, whence it was to be raised into a reservoir on the hills south of the city, and thence distributed by iron pipes through the streets. The cost for a present supply, with works of a capacity to furnish more when required, was estimated at four millions of dollars.

In 1854, a new Water Committee was appointed, composed of Messrs. John A. Dayton, R. C. Brainard, D. P. Barnard, F. G. Quevedo, and Samuel Booth. The chairman, Mr. Dayton, had been actively interested in procuring the bill of June 8, 1853, creating the Long Island Water Company, and a new direction was thus given to the investigations of the Committee. Gen. Ward B. Burnett, the engineer of that Company, was employed, and made a report, dated March 13, 1854, which is based on his previous surveys, and on the data procured for the previous Committees. Assisted by Mr. Samuel McElroy, he submitted maps plans and estimates, which were published in quarto form at the close of the year by the Water Committee. The change of location of the conduit and canal line and of the main reservoir, from Mount Prospect to the present site, is the principal change in this plan, and the details are worked out differently. This plan, under an act passed April 7, 1854, amending the act of 1853 in some particulars, was submitted to the electors on the 1st of June following.

The plan proposed, contemplated a supply obtained from all the streams, as far east as Parsonage creek, furnishing twenty-three millions of gallons daily. The ponds were to have dams, and the surplus water they furnished was to be carried in an open canal to the first pond, or one nearest to the city, which was to be used as a receiving and settling reservoir, and thence by closed conduits to a pump-well near Spring creek, where steam-engines, capable of raising twenty millions of gallons daily, were to send it to the main reservoir, capable of containing at least two hundred millions of gallons, and thence to be sent into the city by pipes. The conduit and canal were to be constructed with a capacity of forty millions in case of future extension of the works, and the whole was estimated to cost four millions and a half of dollars.

At the election, 9,105 votes were cast, of which 6,402 were unfavorable to it, thus again defeating the attempt to furnish the city with water.

The well system was again advocated about this time, in a printed pamphlet, by Mr. James Walters, one of our most respectable citizens, and who had been one of the Water Committee of 1834.

The season having proved a very dry one, no rain having fallen of any consequence for about six weeks up to the 9th of September, gaugings were made of the various streams proposed to be used, and it was found that the long drought had not affected their flow, thus proving their capacity as a source of supply.

Nothing further, however, was done this year beyond the publication of the Committee's report in full, with plates and estimates, as above mentioned.

In May, 1854, some persons having still a preference for the supply from wells, Mr. Dayton procured from the late John S. Stoddard—a graduate of West Point, and who had laid out the streets in the city of Brooklyn in 1835 to 1839, under the act of April 23, 1835—a report on this subject. In it he carefully examines the capacity of wells sunk for this especial purpose to some depth in the main spring. He shows that an area of 29.10 square miles would supply one million of gallons only per day, and that the available area of drainage would not be sufficient to supply the required amount. The streams draining this main spring by a natural process are therefore, as he states, the proper source of a supply. This report seems to have finally disposed of the well system.

On the 16th of March, 1854, Messrs. Henry S. Welles & Co., contractors, made a proposal to the Water Committee, offering to construct the works according to the plans last described, including the purchase of all the ponds and the reservoir site on Cypress hills, and guaranteeing a daily supply of twenty millions of gallons, with works capable of carrying double that amount of water, for four millions one hundred and seventy-five thousand dollars. This proposal included four ponds and forty-eight acres for the reservoir, held by the Long Island Water Works Company, which seemed unwilling to part with this property, so necessary to the construction of the works, and therefore the Water Committee did not entertain the proposal made by Welles & Co.

Another proposal was to construct water works on a plan which embraced the main features of the one designed by Mr. McAlpine, but the canal was to be throughout an open one, and arranged so as to take in all the fresh water that now finds its way to the bays. This with the use of one stream only, the one nearest to the city, and now called Jamaica reservoir, with pump-well, steam power to supply five millions of gallons in twelve hours, and eight miles of mains, was to cost about four millions of dollars. This was presented by Messrs. Joseph Battin, Silas Ford and Henry Ruggles, and was ordered to be printed October 2, 1854, on the recommendation of the Special Committee, Messrs. R. C. Brainard, Peter Wyckoff, and H. N. Holt, to which it had been referred. The association proposed to obtain a charter, with the necessary powers, and to take three millions of the stock, upon the condition that the city should take one million. The Company was to be managed by a Board of Directors, of which the city was to appoint one-fourth. The rates charged for water were not to be higher than those charged in 1842, in New York, and the water was to be supplied to the fire-plugs without charge to the city.

This proposal was not favorably considered, but the associates obtained a charter in 1855, by an act passed April 12th, 1855, incorporating the Nassau Water Company, which gave it ample powers to construct works and supply the city. The city was authorized to subscribe for one million three hundred thousand dollars of the stock.

The Company did no more than was required by law to make its charter effective; that is, the directors named in the act, subscribed each for a few shares, and then proposed to the authorities that the city should subscribe for the amount it was empowered to take.

This offer was made, but was for a long time so ill received by the Common Council, that it was not until September, 1855, that a report was made by the Water Committee upon the subject.

The Water Committee of 1855, Messrs Benj. F. Wardwell, Geo. L. Bennet, J. V. Bergen, E. S. Blanck, and C. C. Fowler, found matters at a stand-still. The Hon. George Hall, elected mayor of the city for the second time, refers, in his address of January 1st, to the difficulty of devising an acceptable plan for a water supply, and recommends that an entirely independent commission should be created to carry out this all-important work.

The city had obtained a new charter in 1854, which went into effect on the 1st of January of this year. It incorporated the village of Williamsburgh and the town of Bushwick as part of the city of Brooklyn, thus increasing the population and adding to the taxable property. With this additional power to bear a public burthen, the cost of water-works suitable for the supply of a population of about two hundred thousand souls, then included in the extended city, would be lightly felt. The debt of the consolidated city at this time was but a little over one million of dollars, with assets enough to balance it, while the taxable property was returned, in 1854, as eighty-eight millions nine hundred and twenty-three thousand dollars. To these figures, taken from the above address of his Honor the Mayor, the following may be added as showing what the state of the city was at this time:

| | |
|---|---|
| Superficial area of the consolidated city | 16,000 acres. |
| Water front | 8½ miles. |
| Inland bounds | 13½ miles. |
| Greatest length and breadth | 7¾ and 5 " |
| Number of buildings (Brooklyn only) | 19,576 |
| Churches | 113 |
| Public schools | 27, with 317 teachers and 30,500 scholars. |
| Numbers of steam ferries connecting with New York | 13 |
| Horse railroads | 30 miles. |
| Two gas companies with pipes laid | 95 " |
| Number of gas-lamps | 2,609 |
| Number of public cisterns | 157 |
| Number of wells and pumps | 547 |

Nine banks, four savings institutions, and eight insurance companies, were in the city. Five daily and two weekly papers were published here. Numerous ship-yards, and manufactories of all kinds, which according to the census of 1855, had turned out twenty-two millions seven hundred and three thousand two hundred and ninety-three dollars' worth of products, were at work, and the prosperity of the city was increasing rapidly.

These figures may be compared with its state only ten years afterwards to prove its rapid growth, and the wisdom of at once inaugurating an effective water supply. The wells were becoming tainted, and large fires could not be subdued with the slender supply afforded by the public cisterns.

Under the name of "The Citizens' Vigilance Water Committee," some gentlemen, in 1854, had met, and appointed the Hon. John Dikeman as chairman of the organization. Messrs. A. J. Spooner, Peter Wyckoff, E. Beers, John E. Cammeyer, C. C. Smith, Barnett Johnson, Rufus R. Belknap, Daniel Maujer and others, appear among its members, but the questions it discussed had been settled already. There was no doubt as to the proper source of a supply, the only obstacle which prevented the beginning of the work being how to get rid of the Nassau Water Company's charter and let the city have the sole control of the enterprise ; but this the Committee did not consider or could not remove.

As above stated, the memorial of the Nassau Water Comany was reported on by the Water Committee in September, 1855, and this time favorably ; but nothing was done until November 15th, when the Common Council, at a special meeting, decided to subscribe for one million of the capital stock of the Nassau Water Company, but not until the charter of the Company had been amended so as to give the city a proper representation in the Board of Directors, nor before two millions had been subscribed and partly paid in by the stockholders. Should the Company commence works within six months after its organization, it was authorized to lay mains and distribute water under regulations prepared by the authorities of the city. The Mayor approved this resolution on the 23d, but the Company did not procure the required amendments to its charter nor fulfil the conditions asked for by the city.

In his annual message, at the opening of the year 1856, His Honor Geo. Hall, the Mayor, dwelt at some length upon the necessity of a supply of water, and upon the present state of the water question ; but nothing could be done by the Water Committee for this year as long as the charter of the Nassau Water Company was in force. The Committee consisted of Messrs. Geo. L. Bennett, Ed. T. Lowber, Geo. M. Troutman, Rd. H. Huntley, and John V. Bergen.

However, a few public-spirited citizens, having made themselves thoroughly familiar with the question, and having, by consultation with engineers, satisfied themselves that the source of supply recommended by the previous Water Committees was the only proper one, sent to the Common Council, on the 14th of April, a communication asking for a conference on the subject. This led to tangible results, for the citizens were wearied with the technical difficulties and the unnecessary delays which had been raised or caused by parties looking less at the interests of the public than at the opportunities of private gain, which the disbursing of large sums might throw into their way. The Water Committee, anxious to accommodate the public, held the proposed conference with the citizens, among whom we name the following: Messrs. Fisher Howe, Conklin Brush, John H. Baker, Abm. B. Baylis, John H. Prentice, Wm. Wall, G. H. Howland, Roswell Graves, etc.

The report agreed upon at this conference was submitted to the Common Council on the 5th of May, by the Committee. After briefly rehearsing the case, the report proceeds:

"The Committee learned that a plan for Water Works had been carefully matured by the Nassau Water Company, and that a contract upon detailed and elaborate specifications and estimates could be entered into with H. S. Welles & Co., for the construction of those works, at a cost not to exceed four millions two hundred thousand dollars, and if certain suggested modifications were made, at a cost of considerable less than four millions of dollars.

"At that sum, the contractors were ready to guarantee a minimum supply of ten millions of gallons per day, within two years, and an additional ten millions of gallons per day within one year thereafter—and to deliver the works in perfect order—including land, water, canal, conduit, reservoirs, engines, &c., &c., discharged of all claims for land, water, and other damages, and all cost in the detail of construction; in short, to invest the Company with the works unencumbered, and without any cost or charge of any description beyond the sum stipulated.

"The plan and the proposed contract had been carefully scrutinized by the eminent practical talents of the gentlemen composing the Citizens' Committee, and their opinion was unanimous, that the plan and contract were highly advantageous and the best that seemed likely ever to be obtained.

"Your Committee, with the lights furnished by the elaborate reports heretofore made by engineers and committees to this Common Council, and the additional data in possession of the Nassau Water Company and citizens, have devoted much time and labor to the consideration of the leading points involved, and particularly to

"The sources of supply;

"The quantity required;

"The best mode of introducing and distributing the Water;

"The character of the works required, both as to material and style, and modes of construction;

"The cost of the works;

"The precautions to be used to secure works of the description contracted for, and against an increase of the cost beyond the stipulated price upon any contingency.

"Your Committee, after a thorough scrutiny of the specifications submitted by Welles & Co., placed them in the hands of A. W. Craven, Esq., Chief Engineer of the Croton Aqueduct Department of New York; his opinion is annexed as part of this Report. The specifications were modified in conformity with his suggestions.

The contract proffered by Welles & Co., was also placed in the hands of J. M. Van Cott, Esq., and his opinion is annexed as part of this Report. The modifications suggested by him were also adopted.

"Our investigations have resulted in the conviction expressed by the Citizens' Committee, that the plan and contract proposed between the Nassau Water Company and Welles & Co., is as advantageous to the public as any that is likely to be proposed, and more advantageous than any ever before submitted to the Common Council and citizens."

The necessary steps to be taken are then indicated:

"The Committee recommend that the Common Council immediately subscribe the sum of one million three hundred thousand dollars to the stock of the Nassau Water Company.

"It may be asked, What security will the city have that their subscription will be faithfully expended and applied?

"In answer, we reply: The Directors of the Nassau Water Company, with a commendable public spirit, expressed their willingness to retire from the direction, and to have their places filled with seven citizens of known character and probity, in whose hands, as directors and commissioners, the citizens of Brooklyn would be willing to repose so great a trust; and in proof of their sincerity, their resignation has been made, and their places filled by the election of the following gentlemen:

JOHN H. PRENTICE, JAMES CARSON BREVOORT,
WILLIAM WALL, NICHOLAS WYCKOFF,
DANIEL VAN VOORHIS, THOMAS SULLIVAN,
NATHANIEL BRIGGS,

whose names had previously been submitted to and approved by the Joint Committee."

Here it may be well to say, that the change in the direction was made principally because the gentlemen named in the act organizing the Company declined to serve, and partly because it was tacitly agreed that the new Board should act as agents only for the city until a law making them commissioners could be procured. Had the Directors chosen to do so, they could have issued stock for the balance of the capital and have sold it in open market; but it was understood that no more stock was to be issued, and that the works were to be carried on under the Nassau Water Company's charter in trust for the city. This involved the assumption of a contract already made and left the new Board but little discretion in making any changes in the plans, however much they might be required.

In conclusion, the Committee offer the following resolutions for adoption:

"*Resolved*, That the resolution adopted by the Common Council, November 15, 1855, for a conditional subscription of one million of dollars to the stock of the Nassau Water Company, be, and is hereby rescinded.

"*Resolved*, That the city will subscribe the sum of one million three hundred thousand dollars to the stock of the Nassau Water Company, and that the Mayor be hereby authorized and requested to sign the Company's subscription books, to give effect to such subscription when the amendments to the specifications and contract between Welles & Co. and the said Nassau Water Company, proposed and approved by this Common Council, shall have been acceded to by said Welles & Co.

"*Resolved*, That the Water Committee be authorized and instructed to do what is necessary to secure the faithful application of the moneys subscribed by the city to the stock of the Nassau Water Company, and to employ such professional aid in performing that duty, as may be deemed necessary, and also to devise the form of the City Bonds to be issued to pay said subscription, and to fix the rate of interest, and the times when principal and interest shall be payable—such interest not to exceed six per centum, payable semi-annually."

The specifications, drawn up by Mr. McElroy, then are given, and also the contract.

The opinion of Mr. Alfred W. Craven, for many years the Chief Engineer of the Croton Water Works in the city of New York, dated April 22d, is added. He makes several valuable suggestions, which were incorporated in the specifications above mentioned. He remarks, very properly: "It may be further assumed that, in this, as in all other works of the kind, there will be, during its progress, certain features which may require revision and modification. Provision should be made for such changes, for it will always turn out to be the most economical course to have the work well done during its progress, and well fitted to the duties which are expected of it. The contract can easily be drawn to embrace the contingency of changes, without, in my opinion, based upon what is before me, together with my understanding of your arrangement with the contractors, the prospect of any *change* which will materially increase the total cost of the work, as now specified.

"Not being called upon to examine, or report upon the general plan of the work, I have tried to confine myself to the specification, as handed to me. The alterations here suggested, can be made without sufficient additional expense to be a bar to their adoption, if they be approved.

"In conclusion, I repeat that the specifications, handed to me, are so far well considered, that if a contract be drawn, defining by words and *plans*, what is now conveyed in the spirit of the specifications, it will be ample in its provisions and details, and, I think, fair to both parties."

Joshua M. Van Cott, Esq., as counsel to the Citizens' Committee, had examined the contract carefully, and made several amendments therein, which were agreed to by the contractors before its execution.

At a meeting of the Common Council, held on the 4th of June, the specifications and contract, as amended, were considered and adopted. On the 9th of the same month the resolutions were approved by the Mayor, George Hall, and on the 17th he subscribed on behalf of the city to one million three hundred thousand dollars of the Nassau Water Company's stock, at the office of the Nassau Insurance Company, in the presence of a number of citizens interested in the prosperity of the city.

Thus the obstacles, which had so long delayed the active inception of this necessary work, were finally removed. Perhaps this delay was, on the whole, of advantage to the works as they now stand. Since 1849, when the question was first seriously examined into, the city had grown with rapid strides, and works as then proposed would have been totally unfitted to supply the quantity of water required now. Time had also been gained to mature the details most carefully, and to each of the Engineers above mentioned belongs a part of the perfection which the Brooklyn Water Works now possess. As they are now built, additional supplies can be passed through them without any further expense than is required to reach the streams to be taken in, and to add lifting power to that already in action. The conduit is large enough to pass over forty millions of gallons daily, and the engine-houses of sufficient size to contain four engines, each capable of raising ten millions in sixteen hours.

The Board of Directors thus appointed met on the 20th of May, and was organized for business by the appointment of John H. Prentice as President, aud J. Carson Brevoort as Secretary. On the 10th of June the resignations of Samuel McElroy as Chief Engineer and of H. C. Murphy as Counsellor to the late Board, were acccepted and a committee was appointed to select a Chief Engineer. On the 24th this committee recommended Mr. James P. Kirkwood, a civil engineer of great experience in his profession, as Chief Engineer, and he was at once appointed to this post. Mr. J. M. Van Cott was nominated as Counsellor to the Board. On the 27th letters were read which had been received from these gentlemen accepting these appointments. The Chief Engineer was directed to commence work at once, by organizing a staff of assistants and proceeding to make the necessary surveys. The Secretary was directed, at the same meeting, to notify the Water Committee of the Common Council, that the Board was ready to commence work. On the 30th, the Chief Engineer and Secretary visited the line of the proposed works for the first time, and within a few days rooms were secured for the accommodation of the engineers and the Board.

On the 3d of July the Water Committee, in answer to the communication of the 27th of June, recommended that the contractors be notified to proceed forthwith to construct the works under their contract, upon entering into the following agreement, viz.: to receive the first $500,000 accruing upon their contract in the six per cent. Water Bonds of the city at par, and the remaining $800,000 in cash, or in bonds at par, according to the election of the Board, in further payment for work. This agreement was duly executed on the 7th, when a call was that day made upon the city for a payment of the sum first named, the bonds to remain in the hands of the Treasurer of the city, and to be issued on the presentation of drafts made by the Board, and signed by its President, Secretary, and Chairman of the Finance Committee. The contractors, at the same time, were notified, according to Section 8 of the contract, that adequate means had been provided for payment of work performed under this contract.

On the 21st the Board resolved to fix upon Thursday the 31st of July for an inaugural celebration of the work, to be held on the site of the principal or Ridgewood reservoir. A Committee of the Board was appointed to make the arrangements for the occasion, and another committee to procure speakers.

The weather was propitious, and the celebration was held at the time and place named. A large concourse of citizens and strangers had gathered together on the hill where the reservoir now is placed and where awnings, platforms, and refreshments, had been provided, all tastefully decorated for the occasion. His Honor Mayor Hall spoke first, followed by the Rev. Dr. Geo. W. Bethune and others.

On the 4th of August the first monthly estimate was received by the Board, from the Chief Engineer, being for the cost of the reservoir site. A commission had been appointed by the *Supreme Court, on the application of* the Board, to estimate the value of all lands required for the works, and of damage to others not taken, which continued to perform this duty until all were satisfied.

On the 17th of September, some changes in the construction of the principal reservoir, and in the location of the pump-well and line of conduit, were recommended by the Chief Engineer, and were accepted by the contractors.

On the 13th of November, Alfred W. Craven, Esq., Chief Engineer of the Croton Water Works, was appointed Consulting Engineer of the Brooklyn Works.

The works had been actively progressing during the rest of the year 1856, and the gauging of the streams had been kept up. No engineering difficulties had been met with, and the Board, with its engineers, had been doing their utmost to push matters actively on in the interest of the city.

At the approach of the legislative session of 1857, a law had been prepared, under the authority of which the city was to become absolutely vested in all the contracts, property, and rights of the Nassau Water Company, and under which the seven Directors were to become a Board of Commissioners to construct the works. All the acts of the city relating to its subscription to the stock of the Company, etc., were confirmed, and the further issue of bonds to the amount of $2,900,000 was authorized. A sinking fund of $50,000 a year was provided for, and the works were pledged for the payment of the whole debt incurred in their construction.

This law, a copy of which is annexed, was passed February 11, 1857. On the 11th of April, 1857, the necessary consent was executed, and on the 22d of June, the Common Council accepted the proposed transfer authorized by the new law. On the 2d of July the transfer of the twenty shares standing in the name of the Directors of the Nassau Water Company was signed, and the Board was organized as a "*Board of Water Commissioners of the City of Brooklyn*," on the 9th. The same officers and engineers were chosen, and notice of the action taken was, on the 13th, directed to be sent to the Mayor and Common Council.

On the 21st of March, the Chief Engineer submitted a report on the subject of adopting a brick conduit in place of an open canal east of Jamaica pond, but the Board concluded to proceed with the construction of the canal, preferring not to alter the original contract.

The financial crisis of 1857 affected the contractors very seriously, for they could not dispose of the bonds at par, and meeting with heavy loss on this account, they asked, in July, to be released from their engagement to take eight hundred thousand dollars of City Bonds in addition to the first five hundred thousand dollars which had been paid to them for work up to that time. Several attempts were made to negotiate a loan on the bonds for the contractors, which failed; but the Board, by the issue of certificates payable at short dates, and bearing seven per cent. interest, succeeded in procuring sufficient means to relieve the contractors without prejudice to the agreement made with them.

Fearing to undertake the cutting of a canal through a continuous stratum of sand, the Chief Engineer, in August, again urged the change he had before proposed, and the Board, concurring in his views, sent a communication to the Common Council, with estimates of the additional cost of such change; but no attention was paid to it. In November, the Board ordered the canal to be commenced. It was found, as predicted by the Chief Engineer, to be totally incapable of conducting the water, and its section was altered with every storm of wind. In April, 1858, the Common Council, having become satisfied that it could not be made efficient, took the initiative towards the desired change, and asked for information; which however being given, no further action was taken in relation to the matter for over a year longer.

All the rest of the work was being actively carried to completion. The pump-well had been completed, as well as the conduit, for five miles to the first pond. The rising mains were in position, and the reservoir in condition to receive water before the autumn of 1858. A great part of the pipe distribution had been completed, and the contractors, by using a small pumping-engine, succeeded in raising water into the Ridgewood reservoir on the 18th of November, 1858, which was Thanksgiving-day. On the 4th of December, the water was let into the mains, and was first used on the 16th of December for extinguishing a fire corner of Myrtle avenue and Schenck street.

The Common Council of 1859 seemed to be even less disposed than the one of 1858, to favor the completion of the great work which was to confer so many benefits upon the city. Soon after its organization in January, an investigating committee was appointed, which, after exciting much public discussion and causing considerable annoyance to the Chief Engineer, satisfied the public, if it did not itself, that all was proceeding well. The slight subsidence of a few stones forming the lining of the main reservoir, caused by the washing out of the gravel through the joints of the stones by the high winds, was talked of by the investigating committee and its friends as if the work were falling to pieces. A few men repaired the damage in less than a week. This and other petty and futile charges were all done away with, and the credit of the Board was not further assailed this year.

In anticipation of a transfer of the works, when completed, to a Board of Permanent Water Commissioners, a law had been prepared, which was passed April 16, 1859. This law also included a section giving to the Commissioners power to expend five hundred thousand dollars more on the works. This was necessary in order to extend the pipe distribution and perfect the open canal.

On the 21st of April, Daniel Van Voorhis, Esq., who had for nearly three years given much of his time and attention to the work, resigned his seat in the Board, and it was filled by selecting Conklin Brush, Esq., who, as Mayor of the city in 1852, had advocated the building of water works, and who had always shown an enlightened zeal in all that concerned the welfare of the city.

On the 28th of April, of the same year, a celebration of the completion of the works was held by the authorities.

On the 14th of May, it was resolved that the water works be called the Nassau Water Works of the City of Brooklyn.

On the 21st of May, the Board accepted a proposal which had been long pending and carefully examined, to change the open canal east of Jamaica pond into a closed conduit at an additional expense of four hundred and fifty thousand dollars, less the amount required to complete the canal. The law of April 16th authorized the city to issue bonds for this purpose, but the comptroller refused for some time to do so, until required to issue them by the Supreme Court of this district.

On the 1st of July the Board appointed Messrs. William B. Lewis and Daniel L. Northup, as permanent Water Commissioners, and the Mayor and Common Council shortly after appointed Messrs. Gamaliel King and John H. Funk, thus completing the new Board according to the provisions of the law of April 16, 1859. Until this time the Constructing Board, in addition to its other labors, had in charge the building of sewers and the distribution of water, duties that required much time and patience to fulfil.

On the 12th of July the transfer of their duties as Sewer Commissioners was made, and the Constructing Board had only to give its attention to the completion and administration of the water works.

In the autumn, much vexatious opposition was again manifested by a portion of the Common Council, which undertook to refuse paying the contractors for the expense hitherto incurred in pumping the water into the reservoir. The matter was finally settled to the satisfaction of the chairman of the Common Council Committee, by the contractors, who had threatened, unless paid, to stop pumping, and also to prevent the Constructing Board from taking possession of the engine-house for that purpose. Besides this, the contractors succeeded in obtaining an advance of $100,000 from the amount reserved out of the monthly payments, an advance which would not have been granted by the Water Board.

His Honor Mayor Powell had done all that lay within his power to forward the work, and on the 2d of November addressed the Board concerning the state of the works; to which a reply was sent in the same day, to the effect that the works were so far completed as to be able to furnish all the water required for the city, and also expressing a wish to transfer them to the Permanent Board as soon as possible, no further work being required beyond the completion of that in hand. The delays which had taken place, be it remarked, were principally caused by the refusal of the Common Council to allow the canal to be altered into a conduit, and by the opposition shown to the regular payment of the contractors, matters which the Constructing Board could not control.

On the 10th of November, the transfer of the whole of the works in use, with reservation of the right of access to the same for their completion under the contract, was made by the Constructing to the Permanent Board.

On the 19th of January 1860, the first of the large pumping-engines was certified complete, having been most carefully tested for capacity and duty.

On the 2d of January, 1862, a certificate was sent to the Common Council of the completion of the whole work excepting Ridgewood Engine No. 2, and the engine-house and engine for the Mount Prospect Reservoir. The second large engine was certified complete after full testing, on the 24th of February, 1862. On the 26th of May the Mount Prospect Engine was certified complete, after trial.

After this date, but little remained to be done by the Constructing Board, beyond settling with the contractors. In July, the Chief Engineer sent in a communication respecting a second pipe main. He recommended that a forty-eight-inch pipe be laid, in order to insure a full head in the city with the increasing consumption of water; and estimated the cost of it at $480,910. After careful consideration, the Board, on the 15th of November, sent a communication to the Common Council on the subject, but no action was taken to authorize the work.

The works had been virtually transferred to the Permanent Board; but one portion of the work still remained under the control of the other Board—the extension of distribution. The law of 1859 had not provided for the transfer of this work, which, however, was really performed by the Engineers of the new Board, though it was necessary to have the different extensions authorized by the old one. All the powers of this last Board were transferred, by an act passed May 11, 1865, to the Permanent Board.

A thorough system of sewerage and drainage being a necessary adjunct of a full water supply, the Board, soon after it had begun work, matured a law authorizing it to inaugurate such a system. This subject has received full consideration in the older European cities, but here little experience has been gathered of any value. However, a law was procured in April, 1857, empowering the Water Commissioners to act as a Board of Sewer Commissioners, with other needed provisions. Julius W. Adams, a civil engineer of much experience, was selected to take charge of this work, which has been steadily progressing since then. By an act, passed April 16, 1859, additional provision was made to carry out the plan, and also for its transfer to the Permanent Board at a proper time.

Pipe sewers, that is, sewers constructed of earthenware pipes, which are extensively used in England, were adopted, and are in successful operation. The small section of these pipes causes them to cleanse themselves, and with all the necessary additions of street basins, man-holes, and outlets, considerable skill and ingenuity is required to plan and project a good system of sewerage. Districts, which are separated by a water-shed, are laid out, and accurate levels having been taken, a sort of under-ground river-basin has to be contrived, with branches and mains, and with a capacity sufficient to carry off the rain-fall and house-drainage.

So much space has been devoted in this work to the description of the Water Works, that but little can be added concerning the Sewerage of Brooklyn. It deserves careful study by those who wish to inaugurate such a public work elsewhere. The Board of Water Commissioners has every reason to congratulate itself upon the choice of the Engineer, who planned the work, and had charge of its extension for several years.

Mr. Kirkwood, the Chief Engineer, closed his connection with the Constructing Board on the 2d of January, 1862, on which occasion the accompanying resolution was adopted.

This paper can scarcely be considered as complete, without some recognition of Mr. Kirkwood's important services in the conduct of this great work to completion, further than that afforded by the tame wording of an official resolution ; but, with the public here and elsewhere, wherever he is known, as with his professional brethren, it would be wholly supererogatory, and to himself distasteful, from the highly eulogistic terms in which such notice must necessarily be embodied.

On motion, the following was unanimously adopted, with the request, that an official copy of the same, signed by the President and Secretary of the Board, be forwarded to Mr. Kirkwood :

"*Resolved*, That the Board of Water Commissioners, in accepting Mr. Kirkwood's resignation of this date, and thus closing their official relations, are desirous of expressing their gratification that the pleasant intercourse heretofore subsisting between them, has not been brought to a termination, save with the triumphant success of this work, under circumstances at times, of great perplexity and embarrassment, and in which the enterprise of its faithful friends and advocates has at no time been more conspicuous, than has been the scientific skill and persevering industry which he has brought to bear, in conducting it to a successful completion.

"(Signed) JOHN H. PRENTICE, *President*.

"JAMES CARSON BREVOORT, *Secretary*."

## LAWS OF STATE OF NEW YORK, IN REFERENCE TO BROOKLYN WATER WORKS AND SEWERS.

Act incorporating Nassau Water Company, April 12, 1855.

Act of February 11, 1857.

Act of April 15, 1857—Sewer Law.

Act of April 16, 1859.

Act of April 16, 1859—to amend Sewer Law.

Act of May 11, 1865.

# REPORT

DESCRIPTIVE OF THE CONSTRUCTION

OF THE

# BROOKLYN WATER WORKS.

BY

JAMES P. KIRKWOOD,

CHIEF ENGINEER.

# BROOKLYN WATER WORKS.

In accordance with the desire of the Board of Water Commissioners, the following description of the specific works of the Brooklyn water scheme has been prepared.

I have not intended to limit myself to such a merely general description as would satisfy the inhabitants of Brooklyn, already more or less familiar with the leading features of the works, but have desired to make it intelligible likewise to all others, who, from whatever reason, may be interested in the details and character of the constructions, as well as in their general cost and results.

Long Island, on which the city of Brooklyn is situated, extends from Brooklyn, which is its western extremity, E. N. E., one hundred and ten miles, with a varying width of from ten to twenty miles.

Its watershed consists of an irregular ridge of low hills, running from the bay of New York to the eastern extremity of the island at Montauk Point. The highest points vary from two hundred to three hundred feet in height. It is mainly constituted of coarse alluvial earth and boulders, running sometimes into blue clay, rarely without clay, and sometimes into a very coarse gravel. The number of small ponds on the ridge evidence the compactness of the material of which it is composed. Its boulders present specimens of a great variety of rocks, many of which are recognizable as similar to the known rocks of Connecticut and of the Palisades on the Hudson river, while many others must have been derived and transported there from rocks with whose positions we are not at present acquainted. The slopes and spurs of this ridge or watershed run into Long Island sound on the north, making an irregular shore there, broken into convenient bays and low headlands. On its south side, towards the Atlantic, its slopes lose themselves in a gravel plain or prairie, which last slopes gently towards the coast. This prairie, whose widest part is called the Hempstead Plains, interposes a distance of from five to fifteen miles between the foot of the hill-slopes referred to and the Atlantic shore, which is very regular in its outer beach; but an inner and more irregular beach exists, formed by the shallow waters of Jamaica bay and Hempstead bay; in other words, the Atlantic shore does not anywhere touch the slopes of this ridge, but is separated from it by the wide gravel plain referred to.

It is from this plain that the Brooklyn water supply is derived. The ridge aforesaid intervenes between this plain and the city of Brooklyn, and has therefore to be surmounted to reach the city. The point of crossing this ridge, which is here one hundred and seventy feet above tide, has been taken advantage of to form a large reservoir controlling the city as regards the head of water, and where a liberal reserve of water can be kept constantly available to meet contingencies and repairs.

Numerous small brooks, originating on the south slopes of the main ridge, cross the gravel plain referred to, delivering their waters into the Atlantic. On the largest of these brooks grist-mills have been long established with corresponding ponds of from eight to forty acres of water-surface, and from five to nine feet depth of water; the fall of water available at their dams rarely exceeds eight feet. It is from the largest of these brooks that the supply of water for the city is derived. Five of the old ponds have been appropriated for this purpose, cleaned out, and converted to meet the new application of their waters, while a sixth pond has been established on a small spring-brook, upon which previously no pond existed.

It is important to understand the relative position of the sources of supply, because there are but few instances of large supplies of water for city use drawn so near to the spring-head, so to say, and which pass through so short a distance of exposure to contamination of any kind before reaching the individuals who use it; and, it may be added, where any vegetable or accidental causes of contamination are more within the control and correction of the water authorities. The brooks are all of them mainly fed by springs delivering directly into their ponds and channels, and the lengths of their watercourses, from where the water is taken for the city, to the summer source in each case, will rarely reach four miles. I will recur again to the character of the water, and to the drainage area of the streams in present use.

It becomes me to mention here the engineers from whose explorations, surveys, and designs, the works to be described eventuated. These were—

| | |
|---|---|
| Major D. B. Douglas | 1835 |
| John B. Jervis | 1851 |
| William J. McAlpine | 1852 |
| General Ward B. Burnett | 1854 |
| Samuel McElroy | 1855 |

Through their united labors, the water-basin and its availabilities, together with the governing conditions of any scheme connected with it, became well understood. The labor of Mr. McAlpine very thoroughly sifted the subject, though his plans were modified by the next in order, Gen. Ward B. Burnett. The specifications and limitations, on which the final contract was based, were prepared by Mr. Samuel McElroy, founded evidently on the survey and report of Gen. Ward B. Burnett. The undersigned was engaged after this contract was closed to superintend its execution, and to furnish the required detail drawings and working specifications in conformity with its provisions. Some few changes were made as the work progressed, with the view of simplifying still further the general plan, to which it is not necessary now to do more than allude.

The date of the contract is 3d May, 1856. The works were commenced on the 31st of July, 1856.

The water was let into the city on the 15th December, 1858, and has since been in uninterrupted use.

The entire works of the contract were not completed, however, and made acceptable to the Water Commissioners, until the 26th May, 1862. These works comprised:

1st. The six supply ponds.

2d. The main conduit and the branch conduits, connecting these ponds with the pump-well.

3d. The pump-well and engine-house.

4th. The two pumping engines.

5th. The force mains, connecting these engines with the Ridgewood reservoir.

6th. The Ridgewood reservoir.

7th. The pipe mains from the reservoir to the city, and the distribution pipes within the city.

8th. The pipe-yard and its accommodations.

9th. The Prospect Hill engine-house and engine, designed to supply a portion of the city situated above the level of the Ridgewood reservoir.

10th. The Prospect Hill reservoir, situated on the highest point of the ground referred to.

I will take the works in the order above mentioned:

## THE SUPPLY PONDS.

The relative positions of the six supply ponds will be best understood by reference to the general map. Their distances from the pump-well, measured along the line of conduit and proper branch, are as follows:

| NAMES. | Distance. | Water area. | Capacity of delivery at the lowest stage of their waters. | | Overflow of each dam above tide. |
|---|---|---|---|---|---|
| | Miles. | Acres. | N. Y. gallons. | Cubic feet. | Feet. |
| Jamaica pond.................. | 5.404 | 40 | 3,275,898 | 419,315 | 7.90 |
| Brookfield pond................ | 7.906 | 8.75 | 2,000,762 | 256,098 | 15.40 |
| Clear Stream pond............. | 8.771 | 1.07 | 784,750 | 100,448 | 11.50 |
| Valley Stream pond............ | 9.708 | 17.78 | 2,541,335 | 325,291 | 12.80 |
| Rockville pond................. | 12.343 | 8 | 2,760,847 | 353,388 | 12.05 |
| Hempstead pond............... | 12.39 | 23.52 | 8,239,947 | 1,054,713 | 10.60 |
| | | | 19,603,539 | 2,509,253 | |

The same streams were gauged by Mr. W. J. McAlpine in October and November, 1851, and the aggregate result then was 23,500,000 gallons. They were again gauged by Mr. Whitlock in October, 1852 (Clear Stream excepted), the result then being 19,514,000 gallons. Add to this our measure of Clear Stream in 1857, and Mr. Whitlock's result for the same six streams becomes 20,298,750 gallons.

I have added, in the above table, the areas of the ponds, their summer deliveries, and their heights above tide. By the summer delivery is here meant the natural delivery of the stream at its lowest stage of water, as ascertained by careful gauging, made during the months of September and October of 1856 and 1857—not including, of course, in this measurement any encroachment upon the reserve of water which each pond, when full, variously retains.

Another stream, Springfield brook, was gauged at the same time. It gave a rate of 676,420 N. Y. gallons at its lowest stage, but is not included in the above table, because it has not been applied to the works, the contractors having intended, and partially made provision, to obtain more than its equivalent from the neighborhood of the pump-well.

The heights above tide always refer to mean high water at the Brooklyn navy-yard, as conventionally fixed at five feet below the coping of the Brooklyn dry dock.

The surface of the water in the pump-well, when there is five feet depth of water in the conduit at its connection there, is six feet and nine inches above tide.

I will confine myself to a description of the Jamaica pond, the work on the others being every way similar in character. The reader will, however, find the drawings given as well for the details of the Hempstead pond, as differing in form somewhat from the others. The plan of each pond owned by the city is besides given, as convenient for general reference.

Jamaica brook, for a certain portion of its course, flows through a low swamp, where there has accumulated a considerable depth of vegetable matter, the growth probably of a long period of time. Jamaica pond, which is supposed to be the oldest artificial pond on the island, rests upon this swamp, and its effect had been to superadd to the material of the swamp a considerable deposit of fine mud, the gathering of some eighty years or more of pondage, derived from the rain freshets of the stream passing over a peaty channel, and from the intermittent action of a mill higher up the brook.

The pond, when first acquired by the water authorities, was overspread by the pond-lily, intermixed with the stems of which, under water, there was found a long slender grass. During the summer months, the large flat leaves of the lily and its flowers covered the entire surface-water of the pond with the exception of its channel-way. The stems of these lilies were frequently six and eight feet in length, and their roots, which were frequently of three inches diameter, penetrated deeply into the boggy deposit below. The grasses referred to were equally long, extending from the bottom of the pond to the surface. These growths are characteristic of every pond on the Atlantic side of Long Island with which I am acquainted, the tidal ponds excepted.

The taste of the water in Jamaica pond was at times sensibly affected by this mass of floating vegetation, and at certain periods of the year its water was faintly colored from the same cause. Its natural water is clear and colorless.

Having drawn off the waters of the pond, the work of its purification from the boggy material aforesaid was then commenced by the removal of the old dam and mill-works, and simultaneously, the removal of all vegetable matter within the entire flowage of the pond and its dam. This material, which, when dry, resembled a light loose peat, was carted outside of the lands belonging to the works, and some of it spread over the neighboring fields with the view of testing its manuring qualities. The depth of this stuff varied from six to eight feet at the centre, running out to nothing on the uplands forming the borders of the pond. The work of removal was troublesome and wet, the material being soft when nearly saturated with water, as much of it was, notwithstanding an abundance of cross-ditching, draining it as much as practicable. Carts were used, and plank-roads laid for them over the slippery stuff; the positions of the plank-ways being changed continually as the process of removal advanced. The bottom, when thus laid bare, proved to be fine sand and gravel of the same character with the material of the surrounding plains.

The bog and sand removed from this pond measured 283,750 cubic yards.

It was in this pond that the remains of a mastodon were found embedded in the bog. Such of the bones as were visible crumbled to pieces soon after exposure, except the teeth, some of which were very perfect. Five of these were known to be found, and probably others were secreted by the workmen. Two of them may be seen in the museum of the Packer Collegiate Institute of Brooklyn, and one in the rooms of the Long Island Historical Society.

The old mill-dam having been entirely removed, the formation of the new dam was proceeded with. This, with the exception of the masonry of the overflow, was composed of earth with a puddle-wall in the centre. The cross section (plate 2) will explain the dimensions of the dam and of its puddle-wall.

The material for the puddle-wall consisted of a mixture of fine gravel and clay. A coarse earthy gravel would have been better, but it was not to be found in that neighborhood. This compost was well mixed and prepared before being placed in the puddle-trench. When so prepared it was laid there in "horizontal layers of from six to eight inches in thickness." Each layer was well worked by spades, and, at the same time, tramped by the feet of the workmen; sufficient water being used to form the whole into a homogeneous water-tight mass. Each layer was allowed a certain time to set, without being allowed to dry, before receiving the next layer.

The amount of clay used to form a proper puddle will always be a matter of experiment and judgment, depending on the qualities of the earth with which it is mixed. In this case the earth was a fine gravel, having no binding properties. The more ordinary earths are already combined with a certain portion of clay, and require, therefore, a proportionally smaller quantity of pure clay to fit them for good puddling material.

The earthwork on either side of the puddle wall was carried up simultaneously. This was composed of sand and fine gravel, the only materials which the place afforded. It was deposited, from carts, in "horizontal layers of about nine inches in thickness" each. A heavy iron roller was repeatedly drawn by an ox-team over each layer, to compress it as much as possible, before the application of the next layer. These banks were carried up to a height of four feet above the surface water of the pond when full; the puddle-wall was run up to three feet above the same surface. On all the earthworks thus formed, the after settlement was very trifling, and in many cases unobservable.

These dams were at first finished with earthen slopes on each side, as required by the contract. A paving of stone was afterwards added on the water side, to defend the bank from the wash of the pond. The paving was nine inches thick, laid upon three inches of small stones, the interstices in the paving filled in afterwards with as coarse gravel as could be procured in that neighborhood.

The land slope and the top surface of the bank were covered with four inches of soil and muck, and then well seeded with grass-seed. The soil, however, of this part of the island is very poor, and the muck, when dry, very light, and liable to be blown off in heavy winds. It will, therefore, require some years of patient attention on the part of those in charge, and some further expenditure annually, to secure a tolerable sod on any of the slopes of the earthworks situated on this plain.

The reader is referred to the plan of the masonwork of the pond (plate 13), for the dimensions of the overfall and sluiceway. The width of the overfall is twenty-one feet; of the sluiceway, five feet. This sluiceway delivers into a circular brick conduit, of forty-two inches interior diameter.

The material of the foundation is sand. A small steam-engine, working a rotary pump, was used to clear the foundation of water. A timber platform—as will be seen on the plan, plate 13—was prepared to receive the foundation course of the masonry. This platform extended, as well, across the position of the apron between the lower wing walls. On the upper side of this platform, touching it, and extending for a short distance into the bank at either end of the masonry, a row of sheet piling was driven, to be sunk not less than twelve feet ("twelve feet or more"), according to the specifications. On the lower side of the platform, across the lower end of the work, the same arrangement of sheet piling was made. This sheet piling was intended to secure the masonwork against undermining, and at the other pond dams it has thus far answered the purpose intended. At this dam, however—I state it more particularly for the benefit of the professional reader—it failed to do so. About two years after the dam and pond had been in use (in July, 1861), it was undermined, the water escaping under the masonry of the overfall, rendering the pond useless for the time being, and damaging the masonry seriously.

During the removal of the damaged masonwork and its foundations, preparatory to the reconstruction of the overfall, the assistant in charge was requested to measure the lengths of the sheet piles taken out, and to lay them aside for examination. There were several pieces found, in the upper row of sheet piling, of but three feet in length. One of these had been carried down stream, the water having broken through where this shallow sheeting occurred. The other pieces of sheet piling were from four to seven feet in length, many of them but four feet. The safety of the work depended on the sheet piling being driven to the proper depth, and the cause of its failure was abundantly evident. The unfaithfulness of the sub-contractor in such a small matter, and equally so of those to whom had been confided the immediate inspection and superintendence of the work, brought about this result.

There had been severe rains through the two previous days, during which the dam of the pond above had given way; the volume of water thus let loose was felt at Jamaica pond, and produced an unusual depth of flow over the stone dam, but its strength was more than sufficient to meet any contingency of this kind, had its foundations been properly attended to.

I resume the description of the masonwork at Jamaica pond:

The timber platform having been duly prepared, a heavy course of granite was laid thereon. The work was then carried up in courses of the same stone, cut to meet the conditions of the plan. The coping stones of the overfall were heavy, and all of the full width. The whole was laid in hydraulic cement mortar.

Two square openings (2ft. × 2ft.) were left in the overfall masonry, as shown in the plan—in addition to the opening through the wing walls for the waste sluice—to allow free passage to the waters of the brook during the construction of the masonwork and its connection with the earthen part of the dam; these openings were afterwards closed.

The buttresses on the inside of the east wing were built to break the masonry lines there, and thereby render the dam more secure against leakage along the back of the wing. The puddle-wall was enlarged as it approached the masonry, and made to cover the whole space between the buttresses.

The paving of the apron was laid in courses with cement mortar. The surface of the paving was placed two feet below the summer water-surface of the brook. At the lower end of the apron a mass of rubble-stone was thrown in to defend the sheet piling from the eddies of the stream.

The stone used was a good quality of granite, obtained from Connecticut. There is no stone found on Long Island suitable for this sort of work.

The mortar was made from fresh burnt hydraulic cement (Rosendale) and clean sharp sand, in the proportion of one of cement to two of sand, mixed dry and then worked up with water. The cement was tested before being used.

The upper face of the overfall, towards the pond, was covered with earth. The reader is referred to the drawings (plate 13) for the form and dimensions of the wasteway sluice and of the conduit sluice. A screen of copper wire cloth was fitted into the masonry in front of the conduit sluice to prevent leaves and fish from entering the conduit. A coarser screen was found necessary afterwards in front of the waste sluiceway, a very small obstruction having sometimes prevented its closing tightly.

2

With the view of avoiding the shallow water which usually prevails around the borders of such ponds, the contract directed the formation of an embankment all round the ponds. This bank was built eight feet wide at top, the top being placed three feet above high water of the pond, with slopes of one and a half to one on each side. The bottom of the water-slope was required to terminate in two feet of water.

A foot-bridge was constructed at the upper end of the pond where the line of this shore-bank crosses Jamaica creek. To pass the small brook which flows into this pond from the east, a culvert was constructed. The forms and dimensions of these passages will be seen in the drawings (plate 18).

Over the conduit sluiceway of each pond, a small brick house is built to protect the working gear of the sluice-gate. (See plate 19.)

I will here give the amounts of muck or bog removed from each pond, with the capacities of water-way, etc.:

| NAMES OF PONDS. | Area of pond ground. | Boggy material removed. | Overfall. | | Waste sluice. | Conduit sluice. |
|---|---|---|---|---|---|---|
| | Acres. | Cubic yards. | Width, feet. | Height of fall, feet. | Size, feet. | Width, feet. |
| Jamaica | 67.33 | 283,750 | 21 | 7½ | 2×3 | .5 |
| Brookfield | 11.88 | 40,797 | 15 | 8.8 | 2×3 | 4.5 |
| Clear Stream | 1.80 | 3,791 | 8 | 3 | 2×3 | 4 |
| Valley Stream | 23.20 | 77,841 | .18 | 8¼ | 2×3 | 5 |
| Rockville | 15.50 | 35,659 | 18 | 8¼ | 2×3 | 4 |
| Hempstead | 26.58 | 107,862 | 24 | 8 | {2×3<br>2×3 | 6 |

I will here state, once for all, that we have no means of giving correctly the prices for which the several pieces of work were executed by the sub-contractors. These prices were not known to the superintending Engineer except occasionally or accidentally. The lump price of $4,200,000, covered the entire works of the original contract, and was the only price known to him officially during the progress of the work. A schedule of conventional prices dependent on this, was necessarily assumed, to admit of the required monthly estimates being made, but as these conventional prices included the profits of the original contractors, they must have exceeded the prices of the sub-contractors, to which, besides, they could have no fixed relation. This system is open to considerable risk and difficulty, but in our case it was carried through without at any time overpaying the contractors, and the required reserve was always at the command of the Commissioners in charge.

## THE CONDUIT.

The contract with H. S. Welles & Co. required the construction of a covered conduit from the pump-well to Jamaica pond only. Beyond Jamaica pond an open canal was specified to be constructed connected by smaller branch canals with the several supply ponds. During the progress of the works authority was procured to dispense with the open canal, and instead thereof, to build a covered conduit, so that now the water from every supply-pond is conveyed to the pump-well by covered passages.

The contract required that Jamaica pond should not be raised, that the conduit should be competent to the delivery of forty million gallons of water (although a supply of but twenty millions was to be provided now), and that the fall from Jamaica pond to the pump-well should be at the rate of six inches to the mile.

These conditions determined the position of the bottom of the conduit, by which I mean the surface of its brick invert, and in effect required that with five feet water in the main conduit at the point where the Jamaica pond branch connected with it, the water derivable from Jamaica brook should be able to flow freely into it. In accordance with this requirement, the bottom of the main conduit at the point referred to is five feet and four inches below the surface-water of Jamaica pond (or the crest of its overfall), equal to 2.57 feet above tide. The position of Hempstead pond determined the fall of the upper or new reach of the conduit. The fall or inclination of the bottom of this portion is one in ten thousand, or six and one third inches to the mile. This fall and the requirement that the conduit should be capable of delivering forty millions of gallons at the pump-well, determined the widths on this upper reach of the main conduit, which, beginning at Hempstead pond with a capacity of flow equal to say 28,500,000 (i. e., the delivery proper to Hempstead pond, 8,239,947 gallons, plus 20,000,000 gallons not as yet collected), increased in size and capacity with the waters received from each pond, until opposite to Jamaica pond, it became of size sufficient to pass forty million New York gallons, as already mentioned.

The width of this conduit at Hempstead pond is eight feet two inches; at the junction of Rockville branch, eight feet eight inches; at the junction of Valley stream branch, nine feet two inches; at the junction of Clear stream branch nine feet four inches; at the junction of the Brookfield branch, nine feet eight inches; and at the junction of the Jamaica pond branch, ten feet.

On the lower reach of the conduit, where it is ten feet wide, its capacity of delivery, with five feet depth of water, is equal to a rate of 47,000,000 New York gallons in twenty-four hours; with two feet eight inches of water, it is equal to 20,000,000 New York gallons in twenty-four hours.

The reader is referred to the accompanying cross sections (plate 23), for the dimensions.

The foundations for the whole length of the conduit were situated below the water lines of the country. The head of water to be contended against varied from two to four feet, except at the lower end, in the heavy excavation next the pump-well, where it amounted to from seven to nine feet. At this place the water was kept down by rotary pumps, worked by steam power, and kept in action during the night as well as during the day. On the other sections hand-pumps or machine pumps were used according to circumstances, and in a few cases the water was kept down by deep ditching toward the bays.

The presence of so much water rendered unusual care and watchfulness necessary in the laying of the foundations. Luckily, we had secured very competent and reliable men for inspectors, men, who with one or two exceptions, were thorough workmen, and who rightly appreciated the responsibility of their positions. To their faithfulness I feel that we are mainly indebted for the excellent condition of the conduit as it now stands. The duties of the Assistant Engineers did not admit of their watching the works, except incidentally, and I do not detract from their merits in expressing my indebtedness to the others.

The sub-contracts averaged each about a mile in length; there was an Inspector attached to every piece of work in progress.

The lower reach of the conduit had been constructed some time before the upper reach was begun. It will be convenient to refer to it first.

This reach extends from the pump-well to the west side of the Jamaica creek valley, 4.848 miles.

At this last point a small terminal basin was formed, arranged to connect with the canal there, and to connect with the Jamaica pond branch conduit. This basin has since been modified to meet the construction of the upper conduit, and covered in.

The lower reach of the conduit is of uniform size throughout. The width is ten feet at the springing of the arch, the side-walls three feet in height, the versed sine of the invert eight inches, and the height of the conduit at the centre eight feet eight inches. The side walls are of stone except an interior lining of four inches of brickwork; the arch is of brick, twelve inches thick, and the invert of brick four inches thick, and on one section eight inches.

The stone used was a good quality of gneiss obtained mostly from Greenwich, Connecticut, and of sufficient size to make good bond. Some stones obtained from the trap boulders of the island were used in the beginning of the work, but they made inferior work and were speedily discarded. At the manholes of the conduit (plate 20) cut granite was used, and the same at the waste weirs and sluices. The bricks used in the conduit were all hard-burnt brick, which had previously been examined and culled of all pale, soft, or defective bricks. The arch, as it was finished, was plastered over on the outside with a thick coat of cement-mortar, and also the outside of each side-wall.

In the preparation of the foundations the contract specifications required a bed of concrete to be first laid fifteen feet wide "where the natural ground is sufficiently compact for the same," and on some short portions of the line where the water could be well drained and kept off, this course was followed; but where the water was more troublesome and not easily kept down, it was found necessary first to lay a platform of plank for the concrete, on which it could be laid and properly packed and set, without incurring the risk of unequal settlement to which it was liable in the wet sand, rendering its removal in that case, and the repetition of the process, necessary. This plank preparation varied in strength in different places, according to the exigencies of the case. It was found a valuable aid towards the rapid progress of the work as well as toward the perfection and security of the foundations. On the section of the work near the pump-well where the greatest amount of water prevailed it was laid as shown in plate 23. The contract thickness of the brick invert was four inches, but on the greater part of this section it was laid eight inches thick.

The depth of cutting on this section was thirty feet. The bottom of the excavation being in water, the tendency of each slope was to slide in, as soon as the excavation reached the water. To prevent this, piles were driven on either side and stout timbers crossed from pile to pile to maintain them in position. Resting against these piles on either side a partition of plank and sheet piling was prepared which held back the sand and admitted of the excavation and pumping going on with but little interruption. This heavy timber shoring had to be maintained here until the conduit was finished and covered.

On one of the middle sections, where a short piece of soft bog occurred, a piled foundation was resorted to. The piles were capped and planked, and upon this plank platform dry stone masonry was carried up to the bottom grade of the conduit.

The concrete was formed from clean broken stone, broken so as to pass through a two-inch ring. Brickbats were also allowed to be broken up and used in the formation of concrete, having been first screened from the brick dust. Two to two and a half measures of the broken stones were well mixed with one measure of hydraulic cement mortar, prepared as already described. The concrete material thus prepared was laid in sections to the required form, well rammed, and then allowed so set firmly before receiving the further work. The centres of the arching were not allowed to be struck until earth had been well packed in behind the side walls, and halfway up the arch. In both cuttings and embankments, the arch was covered with four feet of earth, with a width of eight feet at this height on embankments, and slopes of one and one half horizontal to one perpendicular, on each side. These slopes, and the top surface of the banks, were covered with a layer of such soil as the country afforded, which was not good, and then well seeded with grass seed—a process that will require to be frequently repeated, as already mentioned, before a good sod can be obtained.

The manholes were carried up to the top of the covering bank, and there covered with a heavy slab of cut granite. On the lower reach of the conduit, they were placed on the centre of the arch; but on the upper reach they were placed on the side of the conduit. (See plate 23.)

On a portion of this lower reach, within two thousand feet of the pump-well, thirty small openings (four inches by two inches) were left, at intervals of fifty feet, on the east side-walls, at the springing line of the arch. The sand excavation here varies from twenty to thirty feet in depth, and the underground water-line stands, ordinarily, three, four, and five feet above the level of the holes referred to, the height or head of water being greatest at the holes nearest to the pump-well. A careful experiment, made in February, 1859, after the work was finished, showed a rate of delivery, in twenty-four hours, of 1,402,000 New York gallons from these small inlets exclusively. The depth of water in the conduit on this section averaged three feet eight inches, when the experiment was made; that is to say, it was coincident with the springing line of the arch, and the position of the holes. Had the water stood at five or six feet of height in the conduit, the position below which it is not well to have it when the engines are pumping, the delivery would not have been so great; but in any case, it may be safely taken as worth a daily supply of 750,000 gallons of excellent water when wanted. The inlet holes referred to are closed at present. If the water derivable from this source should be desired hereafter, it will be necessary first to place a protection on the outside of the wall at each hole, which shall prevent any sand flowing into the conduit with the water, and thereby damaging the pumps, as well as producing slides on the excavation slopes.

The conduit terminates in a small arched basin, ten feet wide and fifty-two and a half feet long, placed at right angles to the line of conduit and bearing on the masonry of the pump-well, its arch is twenty-four inches thick. This basin communicates with the pump-well by means of four sluices (plate 31).

The foundation grade of the conduit at the crossing of Spring creek is situated below the water of the creek. The culvert built for the passage of the brook-water is thence liable to be silted up unless watched and kept open by the persons in immediate charge of the conduit works. To facilitate the keeping of such a culvert clear, a trap was in this case built at the upper end of it, into which the silt or sand carried by rain deposits itself. The trap if cleansed out frequently will very much simplify the keeping of the culvert open; if neglected, and hence the culvert allowed to become choked, the waters gathering against the conduit bank might induce serious damage. On the entire line of conduit there are sixteen culverts similarly situated; some in which the water is passed under the conduit through iron pipes, connected with traps as aforesaid. All of these passages want careful watching.

At Spring creek there is a waste-wier whence the conduit water escapes into the creek, when the pumps are not at work. There are also two sluice gates here, for drawing off the water in the conduit, the whole constructed of stone masonry; the covering house is of brick. The reader is referred to the drawings for the form and dimensions of this work (plate 22). At a point $2\frac{186}{1000}$ miles from the pump-well a similar waste-weir with sluice gates is built.

A circular branch conduit, three and a half feet in diameter, connects the east end of this reach with Jamaica pond. This circular branch conduit, like all the branch conduits, was constructed of hard-burnt brick laid in hydraulic mortar.

The water from Jamaica pond was passed through the conduit to the pump-well in November, 1858. It was received into the city in December, 1858, and since that date the city has been supplied with water without interruption, and abundantly.

The upper or new reach of the conduit was commenced in 1859, considerable work having been done previously at this end on the open canal, which to some extent was made available for the conduit works.

While the work on the open canal was in progress, and before the construction of a conduit there was determined on, the contractors, much to their credit, and of their own accord, built circular brick conduits upon the pond branch lines instead of the open canals required by their contract.

This upper reach of the main conduit extending from Jamaica creek valley to its connection with the Hempstead pond is 7.542 miles in length; the entire length of both reaches of the conduit is 12.39 miles.

The materials and workmanship of this reach of the conduit were of the same substantial character as upon the lower reach. The foundations were in this case laid on a planked preparation, with short exceptions. The four-inch brick lining inside of the stone side-walls was omitted, the side-walls being built wholly of stone where stone was used; in some portions the side-walls were built entirely of brick. The brick invert was made eight inches thick throughout, laid on a preparation of concrete resting on the timber platform already mentioned (plate 23). The conduit was covered with the same depth of earthwork, but the slopes of the banks were inclined two to one in this case.

The engineer will perceive, on inspection of the conduit cross-sections, that the conduit does not possess strength enough in side-walls to meet the thrust of the arch, independent of the support received when in use from the earth backing. The conduit is dependent on this backing as its sustaining abutment, and it is for this reason that we have been so particular in requiring the earth to be rammed and packed behind the side-walls before the removal of the centering, and in afterwards continuing this careful packing of the back-filling at least as high as the top of the arch. In the Croton conduit, and still more so in the Cochituate and the Washington conduits, the same condition of things exists. If the earth is unfaithfully applied, or if the conduit leaks after it is applied, so as to produce a settlement or movement of that earth, the arch will in all probability press out the side-walls, or in some way or other show that a corresponding movement in the masonry has taken place. This movement is generally slight, but it is a fertile source of trouble and expense from the difficulty experienced in making a permanent correction of it.

At Jamaica creek, at Valley-stream creek, and at Rockville creek, the arch bridges constructed there (see plates 21, 27, and 28) were arranged to include waste-weirs and two sluice gates at each bridge for drawing off, when necessary, the waters of the conduit. These bridges were built for the canal, which required more space in width than the conduit. They have been made available for the conduit, but at the expense of some clumsiness in superfluous space. Small brick houses were built over each of the bridges to protect the waste-weir and sluice-gates (plate 24). The stone used in these bridges was granite, cut where necessary to meet the designs, and laid in hydraulic cement mortar. The foundations were started from heavy timber platforms, having a row of sheet piling at the upper and at the lower end of each platform. The channelway under the bridge and between its wings was paved with stone laid in hydraulic mortar.

Wherever roads crossed the line of conduit, the crossing was ballasted with broken stone twelve inches in thickness. The arch at these crossings may be seriously damaged if the wheel-ruts are allowed to wear through the stone ballasting and to reach the brickwork.

From all the supply ponds, except Hempstead pond, at which the main conduit terminates, their waters are conveyed to the main conduit through circular brick conduits, as already alluded to. The lengths of these, with their diameters, are as follows:

| NAMES OF PONDS. | Inside diameter. | Length. |
|---|---|---|
| | Inches. | Feet. |
| Rockville | 30 | 1,872 |
| Valley Stream | 30 | 2,103 |
| Clear Stream | 24 | 1,980 |
| Brookfield | 24 | 2,877 |
| Jamaica | 42 | 2,937 |

Upon each branch conduit there is an arrangement near its mouth for cutting it off from the main conduit, to meet repairs.

Wherever water was found in the foundations of these branch conduits, a mould of one-inch plank was laid to the form of the lower semi-diameter, to receive the brickwork. These branch conduits are covered with four feet of earth.

It may as well be mentioned here, that all the lands belonging to the works built under the contract of H. S. Welles & Co., are fenced in. The ponds are fenced with a picket fence, the posts of locust; the main conduit and branches with a post and rail fence; the reservoirs and engine-house grounds with an iron fence, and the pipe-yard with a brick wall in front and a timber fence in rear. The only exception is the strip of land between the Ridgewood engine-house and the Ridgewood reservoir. This piece, three thousand two hundred feet in length and twenty-four feet in width, is not yet fenced. (See plate 42, for drawings of these fences and of the gateways.)

### ENGINE-HOUSE AND PUMP-WELL.

The reader is referred to plates 29, 30, and 31, for the form and dimensions of the engine-house and of the pump-well.

The pump-well connects by four sluiced openings with the terminal basin of the conduit.

The material of the foundation was sand of a very uniform quality. The site having been excavated to a sufficient depth, and the water kept down by pumping engines, a strong timber platform was laid upon it, extending as well over the ground occupied by the conduit terminal basin or pocket, as it has been sometimes called. Upon this platform twelve inches of concrete was laid in layers; each layer being well rammed and allowed to set before the application of another layer. Upon this bedding of concrete the stone work of the pump-well was begun, as well as the preparatory stone work of the pocket. The entire pump-well is built of heavy coursed granite, carefully cut to meet the conditions of the plan, and laid in fresh hydraulic mortar. The invert of the bottom of the pump-well was also built of coursed granite, radiated, as shown in the drawings, and laid in cement mortar.

The bottom of the pump-well, or the surface of the invert which is flat, is situated three inches below tide, being two feet below the bottom of the terminal basin of the conduit. This gives seven feet of water in the pump-well, when there is five feet of water in the conduit. At present (Nov., 1864) with the superabundance of water at command, there is usually eight to nine feet of water in the pump-well when the engines are pumping. Low cross-walls divide the water space of the pump-well into two parts, that the separate pumps of the engines may each be reached for repairs, without interrupting the working of the other. The four sluices already mentioned connect the conduit with each of those parts.

The mason work of this pump-well has proved to be remarkably tight and has shown no indications of any settlement since it was built. The iron sluices have not proved to be as tight as is desirable, but this kind of defect is readily remediable.

The engine-house is of such dimensions, and so placed, as to include within its interior the pump-well, with sufficient space besides to receive four large pumping engines, each having a capacity of delivery of ten millions New York gallons, daily.

Whenever the word gallon is mentioned in this paper, it is always to be understood as the New York gallon, unless otherwise designated. This gallon contains eight pounds of water. One hundred gallons are equal to 12.8 cubic feet, and one cubic foot is equal to 7.8125 gallons, on the assumption that one cubic foot of fresh water weighs 62.5 lbs. avoirdupois.

The interior dimensions of the engine house are eighty-four feet by sixty-nine feet, of each boiler house sixty feet by forty-five feet. The boiler houses form wings to the main building, and are each intended to receive the boilers of two engines. The eastern wing is now occupied by the six boilers of the two engines now in use.

The walls of the engine house and its wings are built of hard burnt brick faced with brown pressed brick. The facings of the doors and windows, the quoins and cut stone work of the chimney, are of fine-dressed Connecticut freestone. The roof consists of iron trusses, sheathed with two layers of one-inch pine boards and covered with tin. The floor consists of yellow pine plank resting on wrought-iron beams. The reader is referred to plates 29 and 30 for the architectural character of the building, and the dimensions of its details. The design was made for me by Julius W. Adams, Civil Engineer. The foundation walls were built of granite and gneiss. The whole of the mason-work, whether of brick or stone, was laid in hydraulic cement mortar. An iron footway runs round the interior of the engine house, situated twenty-five feet above the floor from which the upper parts of the machines can conveniently be examined.

The heavy stone work on which the pumping engines rest, and to which the bed-plates and cylinders of each machine are bolted, is built distinct and separate from the engine-house walls, which do not in any way come in direct contact with the supporting works of the pumping machinery.

The builders and designers of the pumping engines (Woodruff & Beach, of Hartford, by their superintendent, Mr. Wm. Wright), supplied us with the forms and dimensions of foundations desired by them for each engine, and the stone work of these foundations, or engine-walls, was built strictly in accordance with their plan. This portion of the work was built of heavy granite, upon concrete foundations.

Certain portions of these masses were bolted together. The reader is referred to the drawings (plate 32) for its form and mode of connection with the engines. The foundations of the first engine were built while the construction of the engine-house was in progress. The foundations of the second engine were put in after the house was finished and in use, the character of the second engine not having been determined on when the first was under way. The masonry of the foundations of the second engine is of the same massive character as that of the first, consisting of heavy granite work, otherwise secured by heavy bolting. At the engine-house these foundation walls are quite accessible, and can be readily examined by strangers, that floor being lighted with gas for the convenience of the engine attendants.

The chimney is placed at the southeast corner of the eastern boiler-house, but its masonry is not connected with the walls of the boiler-house. The western boiler-house is not now in use, and the chimney for that end remains to be constructed when that boiler-house shall be wanted. The present chimney is one hundred feet in height, and of forty-eight inches interior diameter. The base, for fourteen feet from the ground, is square ; the remainder of the chimney is octagonal, and constructed of the best hard-burnt brick, with an ornamental iron cornice at top. An interior cylinder of brick, separated six inches from the outer walling of the chimney, rises from the foundation to within twenty feet of the chimney top. A horizontal brick flue, of fourteen and one fourth square feet area (height forty-eight inches, by forty-eight inches width), connects the chimney with the several boiler-flues.

## RIDGEWOOD ENGINES.

The two engines are condensing beam engines, possessing some of the peculiarities of the Cornish engine, and intended by the designer (William Wright) as an improvement on that type of engine, both as regards economy of first cost, and economy of working expenses.

The Cornish engine is single-acting ; that is, it receives steam at but one end of its steam-cylinder, and has but one pump ; the Ridgewood engine is double-acting in its steam-cylinder and has two lifting pumps, one at either end of its walking beam, the pump at the steam end being placed in the well immediately underneath the steam-cylinder ; the pump at the other end of the walking beam is connected with the well by a pipe of three feet diameter, the well-pump cylinder being made in effect part of this pipe.

The pump-pistons in descending pass through water in motion ; in other words, while the piston of the lower pump is rising, its water reaches the rising main by passing through the upper pump, whose piston with its valves open, is then making its descent, and *vice versa.* Lest this peculiarity should prejudice the working of the machine, an outer cylinder surrounds the pump-cylinder proper, leaving an annular space of seven and one half inches in width, equivalent to an area of 7.6 square feet, in addition to the area (7.06 square feet) of the pump-cylinder.

The arrangement secures a very free delivery of the water from each pump-chamber.

The reader is referred to the drawing (frontispiece) to understand the peculiarities of this engine.

In the Cornish pumping-engine the steam does not act immediately upon the water, but lifts a weight (a cast-iron column of ten feet, more or less, in length), and this weight is the motor which directly acts upon the column of water. In the Ridgewood engine the steam, in effect, acts directly upon the water, moving instead of a short column of cast-iron, a long column of water, in the present case three thousand four hundred and fifty feet in length combined with one hundred and sixty-three feet of vertical height.

It is not found in practice convenient or safe to start this long body of water as rapidly as the short column of iron can be started in the case of the Cornish engine, nor, on the other hand, can the Ridgewood engine afford to start its column of water as slowly as the Cornish cast-iron weight starts it. The steam-motor must be comparatively rapid through the first half of its stroke, the iron motor or plunger is most rapid towards the latter half of the stroke. In the Ridgewood engines the large air-chamber very much ameliorates the conditions indicated.

A direct, so to say, instead of an indirect action of the steam on the column of water to be moved, forms the chief distinction of the Ridgewood engine as compared with the Cornish engine.

The length of stroke is in neither engine limited by a crank, nor is the power in either engine, to any extent accumulated from stroke to stroke by a fly or other contrivance. At each stroke of the engine the whole steam used is spent, nothing is transferred to the next stroke, unless it may be said of the Cornish engine, that the cushioning of the steam forms a slight reserve towards the next movement.

The stroke of neither class of engine being limited by a crank, its length is variable and dependent upon the variations of pressure in the boilers, and upon the manipulation of the steam-valves by the attendant. This variation is generally slight on the Ridgewood engines; on Cornish engines the full stroke is frequently shortened as much as six inches. The stroke of the piston cannot, however, overreach the full stroke so as to strike the cylinder-head or bottom without first destroying a certain arrangement of beams placed to defend the cylinder from such a contingency.

This peculiarity in these two types is at once a great advantage and a disadvantage. It renders the engine so sensitive in its action as to require more than usual faithfulness on the part of the attendants, both at the engine and at the boilers. If the work of firing the boilers is not so managed as to maintain great uniformity in the pressure of steam there, the variations will become immediately perceptible upon the machine by corresponding variations in the length of stroke, and the attention of the engineer in charge will be kept nervously on the stretch to meet the carelessness of his firemen. This close attention can never safely flag for a moment, and hence it requires a very superior class of men to be in charge of so sensitive a machine.

This is its great advantage as regards economy of fuel. On the other hand, this sensitiveness, notwithstanding the best attention of the engineer, will sometimes bring about an overreaching of the stroke and greater or less damage to the engine. This is its disadvantage, a disadvantage which is not felt in a crank-engine in this way where considerable variation in the steam used may occur, without other effect on the machine than an increase in the rapidity of its motion.

A Cornish engine sufficiently large to perform the work of one of the Ridgewood engines, would have had a cylinder of at least one hundred and thirty inches in diameter. The cylinder of the engine referred to is eighty-five inches in diameter, Our engine builders were all very shy of undertaking to build so large an engine of the Cornish class, and our Water Commissioners, on the other hand, were not willing to vary in this respect from the contract requirements as to the daily work of the engine.

To build so large a Cornish engine in the United States would have been in some degree an experiment; the tendency of machinists being usually to economize too much in material, in all cases where their own first notions have not been corrected by actual use or by long practice. I have referred thus much to what is called the Cornish engine as being a standard in great favor with most engineers in connection with water supply, and as having been evidently (though the name is not mentioned in words) the kind of engine which was contemplated under our contract with H. S. Welles & Co.

Neither of the engines now in use, as first constructed, fulfilled the expectations of the builders, but in each case certain alterations and improvements, small but important, had to be made before the engines could be accepted. Such will always be the case with new machines, whether the novelty consists in arrangement or dimensions. A Cornish engine would have been in every respect so much larger and more massive than either of these machines, that the delays in perfecting it would have been proportionately tedious, and in that case we may feel assured that the city would not have been accommodated with the Ridgewood water quite as early as under the existing arrangement.

The steam-cylinder of engine No. 1 is ninety inches in diameter, of No. 2 eighty-five inches in diameter ; the full stroke in each case is ten feet.

The two pumps of No. 1 are each of thirty-six inches diameter, with the same length of stroke as the engine. The two pumps of engine No. 2 are each of thirty-six inches diameter, with the same length of stroke as that engine ; the pumps of each engine deliver into an air-chamber of seventy-eight and a half inches interior diameter and six hundred and eleven cubic feet capacity ; thence the waters pass into the rising main which starts from this air-chamber. A lifting-valve inside of this air-chamber separates it from the pump, opening and closing with each stroke of the pump. The air-chamber is of cast-iron in three pieces, flanged and strongly bolted together at the flanges—the thickness of the iron is one and a half inches.

There are three boilers to each engine, each boiler being thirty feet in length and eight feet in diameter. The heated gases of the fire-chamber are passed through large flues to the end of the boiler, returned through smaller flues to within six feet of its furnace, and then passed underneath to the large brick flue which connects with the chimney. There are two large steam-pipes, one for each engine, so arranged that they can be connected with any of the six boilers. The drawings (plate 33) will show the dimensions and forms of the several parts of the boiler. The engines are usually worked under a pressure of eighteen pounds of steam in the boilers.

The contract required that each of the Ridgewood engines, when otherwise fit for acceptance, should be subjected to a working trial, during which it should show an ability to deliver into the Ridgewood reservoir ten millions New York gallons in sixteen hours, and for "duty," so called, should at the same time be "tested by the standard of not less than six hundred thousand pounds of water raised one foot high with one pound of coal."

The force-main in each case is of thirty-six inches diameter and three thousand four hundred and fifty feet in length. The water has to be lifted one hundred and sixty-four feet.

These trials were very carefully made under the direction of competent mechanical engineers.

Engine No. 1 showed a capacity of delivery of 10,293,102 New York gallons in sixteen hours, and a "duty" equal to 607,982 pounds of water raised one foot by one pound of coal. Engine No. 2 showed a capacity of delivery of 10,632,366 New York gallons in sixteen hours, and a "duty" of 619,037 pounds.

The details of these trials are given in the Appendix.

The engines, as in all similar cases, were placed in their best working condition for these trials by the constructors, and the boilers were fired with great care and with the best quality of the kind of fuel contemplated to be used thereafter. Under the ordinary every-day working of the engines the same high condition of the machine cannot be maintained; the wear and tear and contingencies of the service affect the results then, and the same economy of fuel is never attainable.

Before the plan of engine now in use here was decided on, and contracted for, certain experiments were made on the pumping-engines of the Jersey City, of the Hartford, and of the Cambridge water-works, the results of which it will be in place to mention here, the reports of the mechanical engineers who conducted these experiments being now out of print.

During the early stage of our work (1857) it was contended by the machinists of the Hartford pumping-engine, and by intelligent parties interested in the construction of an engine upon the same general plan at Detroit, that the Hartford type of engine was capable of a much higher economy in fuel or "duty" than the Cornish form of pumping-engine, which last was considered by some of our best authorities here and by high authorities in England to be the best and most economical form of pumping-engine in use. Certain engines in the neighborhood of London, built by Mr. Simpson, had done as well or better, but their reputation was not then as well established.

To ascertain whether the Hartford type of engine was worthy of the very high position which its friends claimed for it, the commissioners authorized Mr. J. C. Brevoort and myself to have an experiment made on the Hartford engine and also on the Belleville engine of the Jersey City water-works, which was the largest Cornish engine in the country, with a view to ascertain their separate and comparative efficiencies as regards economy of fuel consumed. To this end we procured the assistance of two mechanical engineers of well-established experience and reputation, Mr. Charles W. Copeland and Mr. W. E. Worthen, and left them to conduct the experiments according to their best judgment.

An experiment was afterwards made, from similar motives, to ascertain the comparative efficiency of the Cambridge type of double-cylinder pumping-engine, built by Mr. H. R. Worthington.

The officers of the corporations owning these engines afforded us every facility within their power to enable us to conduct the respective trials satisfactorily. The constructors of the engines were equally serviceable and courteous, and in each case prepared their engine for the trial by placing it in its best working order, and selecting for the occasion their own engineers and firemen.

Inasmuch as such experiments are not very often made, and are of much interest to engineers, I have thought the details of these worthy of record in the Appendix.

The "duty" results, that is to say, the work done in each case reduced to the conventional form of the pounds of water raised one foot high by one pound of coal, were as follows:

| Location of Engine. | Date of Experiment. | Duty per 100lbs. Coal as taken by Indicator. | Capacity of Pump by Measure. | Capacity by Discharge. | Lift. | | OBSERVERS. |
|---|---|---|---|---|---|---|---|
| | | | | | Absolute. | Indicated. | |
| | | | Cub. feet. | Cub. feet. | Feet. | | |
| No. 1. HARTFORD............ | Dec. 5, 1856 | 55,068,516 | 20.532 | 19.665 | 119.063 | 139.651 | Messrs. Copeland & Worthen. |
| " 2. " ............ | " 6, 1856 | 62,661,715 | " | 19.259 | 118.159 | 137.678 | " " " |
| " 3. BELLEVILLE .......... | " 19, 1856 | 72,115,396 | 72.051 | 65.465 | 159.824 | 166.834 | " " " |
| " 4. " .......... | " 20, 1856 | 69,533,221 | 71.827 | 66.058 | 159.087 | ...... | " " " |
| " 5. CAMBRIDGE........... | March 30, 1857 | 70,463,750 | ...... | ...... | 72.29 | 83.145 | " Morris & McElroy. |
| " 6. " ........... | June 26, 1857 | 71,278,486 | ...... | ...... | 69.85 | 75.900 | " Graff & Smith. |
| " 7. HARTFORD............ | July 14, 1857 | 65,505,469 | ...... | ...... | 120.5 | 140.116 | " Graff & Worthen. |
| " 8. " ............ | " 15, 1857 | 68,974,549 | ...... | ...... | ...... | 140.139 | " " " |
| BROOKLYN. | | | | | | | |
| " 9. RIDGEWOOD, 1........ | June 12, 13, 1860 | 60,798,200 | 135.935 | 133.632 | 160 | 170 | Messrs. Smith, Graff & Worthen. |
| " 10. " 2........ | Jan. 9, 10, 1862 | 61,903,700 | 134.080 | 130.854 | ...... | ...... | W. E. Worthen. |
| " 11. PROSPECT HILL...... | May 13, 17, 1862 93 hours. | 64,957,700 | ...... | ...... | ...... | ...... | " " " |
| " 12. " ...... | May 16, 17, 1862 10 hours. | 68,404,200 | ...... | ...... | ...... | ...... | " " " |

In the above tabular statement, the "duty" is referred to the measured stroke of the pump and the *indicated* lift, which includes the friction of the main. The actual discharge of the pump is given in the same table, that the reader may be able to appreciate the difference.

The Hartford engine is a double-acting crank and fly-wheel engine, the form and action of the pumps being its speciality. The Cornish engine has been already described. The Cambridge engine is a double-cylinder, double-acting engine. The drawing in the appendix will explain the peculiarities of each.

The second experiment of the Hartford engine was made after certain alterations had been made in the machinery, as mentioned in the proper report. (See Appendix.)

These results, though convenient exponents of a very important branch of the economy of an engine, do not necessarily establish a proportionate superiority or inferiority in the respective machines on all other points. The first cost of the machine must be taken into the comparison, the cost of its maintenance and repairs, the time lost in making repairs, and the expense of attendance. A high "duty" is an important evidence of economy as regards fuel; but this may be accompanied by want of simplicity or want of right proportion in the parts, elements upon which the life of the machine and its capacity for uninterrupted daily work essentially depend.

The Ridgewood engines have thus far performed their work well, and with less delay and repair, we should judge, than has occurred during the first few years of most of the pumping-engines in the country. The daily consumption of the city has reached, during the last week of May, 1864, an average of 7,816,159 gallons (1,000,468 cubic feet) in twenty-four hours.

### FORCE MAINS.

There are two rising mains connecting each engine separately with the Ridgewood reservoir. Each main is composed of cast-iron pipes twelve feet in length and of thirty-six inches interior diameter. The length of each main is 3,450 feet; the difference in level, of the two ends, one hundred and fifty-two feet. There are four thicknesses of pipes in each main. The pipes at and in the neighborhood of the engine-house (76) are one inch and a half thick, the next section (77) is reduced to one inch and three eighths in thickness, the next (185) to one inch and a quarter, and the upper section terminating on the reservoir to one inch and an eighth thick. The specified weights of these pipes, from which they were not allowed to vary more than four per cent., were as follows:

| | | | | |
|---|---|---|---|---|
| The pipes | 1½ | inch thick | 5,280 | pounds each. |
| " | 1⅜ | " | 4,870 | " |
| " | 1¼ | " | 4,460 | " |
| " | 1⅛ | " | 3,800 | " |

The sockets, as will be seen by the drawing (plate 43), were six inches deep. No yarn was used in the joints of these pipes, which were fitted the entire depth with lead, and calked with much care, inside and out. Notwithstanding these precautions, a number of slight leakages at the joints occurred during the first year of the pumping, easily corrected in each case by recalking. There has been no leak for the past three years.

These mains, at their upper termini, do not pass through the reservoir embankment, but deliver into a chamber of masonry, their outlets being nearly coincident with the high-water level of the reservoir. This was done that the force mains for their entire lengths might each be accessible for repairs without breaking into the bank of the reservoir or rendering necessary the lowering of its reserved water.

At about the centre of each force main, near the Jamaica road, a check-valve is placed on each main, intended, in case of the failure of a pipe, to reduce the damage which might thereby be occasioned, and especially to keep the escaping water away from the engine-house. The form and size of each valve will be seen at plate 47. The upper chamber of the valve is connected to the lower by a three-inch pipe with a stop-cock on it. By this means the water can be drawn from the upper half of the main when desired. This three-inch pipe has been kept open when the engine is pumping, the experience in some rising mains having shown that without such a precaution a slight collection of air is apt to form immediately below the check-valve, causing the engine to start the next stroke with a jerk.

In the case of the check-valve of the Ridgewood mains, the precaution seems to have been unnecessary for this purpose, as the valves, so far as can be judged by the sound, do not close except when the engine stops work. While it is working, the flow under the effect of the air-chamber is so continuous as not to permit the valves then to come to their seats. These valves are reported to be very tight, and to work very satisfactorily.

At the engine-house end of the mains the check-valves in each air-chamber disconnect the mains from the engines and pumps, and relieve these from all water-pressure when the engines are at rest.

### RIDGEWOOD RESERVOIR.

The reservoir grounds include 48.4 acres of land. They are enclosed by a neat fence of wrought-iron, eight feet high, with a small gate on the south side of the grounds, and an elaborate gateway on the north side, upon the Cypress hill turnpike road. The reservoir occupies about two thirds of the grounds, the unoccupied part is available for similar use hereafter.

The grounds are kept in neat order, but no trees were allowed to be planted there, because their leaves would be blown into the water, and besides, to some extent, detracting from its purity, might clog the gates and screens of the outlet-chamber. For similar reasons, it was judged best not to allow carriages or horses to use the banks of the reservoir as a driving-path, the dust from which, as well as the manure of the horses, would, in that case, find its way into the reservoir.

The reservoir is divided into two compartments; the water area of the eastern division is 11.85 acres, and of the western division 13.73 acres; in all 25.58 acres.

The contract required that the reservoir should have a capacity of not less than 150,000,000 of New York gallons when full. Its capacity now, when full, amounts to 161,221,835 gallons, equal to 20,636,395 cubic feet.

The reservoir grounds are situated on the crest of the central, hilly range already mentioned; the original surface was knobby and irregular with a piece of swamp touching the eastern boundary, not an unusual circumstance upon the heights of this range, wherever their very irregular surfaces produce hollows or cups retaining the summit-waters.

The position of the reservoir on the grounds, and the position of its bottom, as regards height above tide, were determined solely by economical considerations, the object being so to place it as that the earthwork of forming it should be a minimum. The contract required it to have a "surface elevation of not less than one hundred and fifty feet above mean tide;" a provision which could be satisfied by any arrangement of the work which the ground admitted of. The surface-water of the reservoir as built, when full, stands one hundred and seventy feet above the tidal base referred to.

The irregularity in the shape of the original surface produced a corresponding irregularity in the depths of the embankments forming the earthen boundary of the reservoir, which vary from thirty feet in height on the east side to four and five feet in height on portions of the west side. The inner slopes of the reservoir rest consequently, partly on artificial embankment and partly on the natural earth of the position, excavated to the necessary form, as will be seen by inspection of the cross-sections of these banks (plate 38).

The first step in the construction consisted in the removal from the entire site of the reservoir, and its banks, of all vegetable soil or vegetable matter of any kind, which was laid aside to be afterwards replaced on the outer slopes of the finished embankments.

The material excavated from the ground in the process of shaping each compartment of the reservoir was used in the construction of its embankments. This material consisted of a stiff, coarse earth, of excellent quality for such work, but full of small stones and boulders. The boulders were broken up and laid aside for paving. The small stones were also carefully removed—no stone exceeding four inches in diameter being allowed to remain in the earth which was applied to the construction of the embankments.

The outer embankments are twenty feet in width at top, which is situated four feet above the high water level of the reservoir; the division embankment is fifteen feet in width at top, which is situated three feet above the same water level; the slopes outside and inside are made at the rate of 1½ horizontal to 1 perpendicular. The situation of the puddling and puddle wall will be best understood by reference to the cross section (plate 38). In embankment the puddle wall is placed in the centre of the bank; it terminates at two feet above the high water line, and at that height it is three feet in width, increasing in width below that point at the rate of one inch to the foot on each side. On those portions of the inner slopes which rest on the natural earth, the puddle is placed upon the slope twenty-four inches in thickness. Where the natural earth ceases, and the artificial earth work begins, it is carried from the face of the slope to the puddle wall already mentioned, situated in the centre of that earth work.

The puddle consisted of a mixture of the earth of the excavations (selected and freed from stones for that purpose), and of a stiff, white clay found in the neighborhood. The amount of clay required to form reliable puddle will vary with the nature of the earth mixed with it, and can only be ascertained by trial and practice. Too much clay is as objectionable as too little. The puddling material used here was very excellent, and the workmen who applied it were of the best description, having had a previous experience and training in this kind of work.

The clay was first well broken up, and thoroughly mixed with the right proportion of earth; it was then carried to its place, whether on the bottom, on the slope, or in the centre of the embankment, and laid in horizontal layers of six inches in thickness; water was then applied to each layer, and the gang of puddlers proceeded to work it with their spades and feet into a stiff, heavy paste, which, when sufficiently worked, was allowed to set before the application of another layer. If exposed too long to the sun, it will over dry and crack, and will then require to be re-worked before the addition of another layer. When the workmen are not well practised in its application it may sometimes be advisable to mix it with water into the proper consistency before depositing it in place, but in that case it should be well re-worked in position.

It is important to its value, as a water-tight defence, that it should always be damp, and in this consists the advantage of placing it in the centre of an embankment wherever it is practicable to do so. The earth work of each piece of embankment was always carried up simultaneously with the puddle wall in horizontal layers of nine inches in thickness. The earth was delivered upon the embankment from carts. Each layer was repeatedly rolled over with a heavy iron roller to compress the earth as much as practicable, and defend the embankments, if possible, from subsequent settlement. The result was remarkably satisfactory, as was shown by the test levels taken at intervals after their completion. The line of the paving on the slopes, of the guard fence on the top of the inner slope, and of the top surface of the banks all round, sufficiently attest this to the eye.

In the construction of these embankments (as should be the case with all important embankments for the retention of water), the work was intended to be done in a manner which should render the bank so compact as to be water-tight independent of the puddle. The material used and its workmanship should always have this in view. The puddle is a wise precautionary addition intended to meet unforeseen defects, and to render the desired result certain.

The entire bottom of each division of the reservoir, after having been graded to the proper lines, was covered with two feet of puddle in the manner already described. The surface of this puddle, when finished, was covered with a thin layer of gravelly earth. The bottom in each case has a fall of eight inches from the south side towards the drainage-pipes of the effluent chamber, to admit of the entire water of either compartment being drawn off when so required for cleaning or repairs.

On the completion of the banks, their slopes were dressed off to the required inclination. The outer slopes were then covered with the reserved soil, and seeded. The inner slopes were paved with dry stone paving, the stone being derived from the boulders of trap-rock found in the excavations and in the vicinity. This pavement was specified to be twelve inches in thickness, laid upon a bed of gravel or small stones and well packed and pinned. The bed was directed to be made six inches thick—as actually laid, the paving was nearer eighteen inches thick at the bottom; on the upper two thirds it was rarely twelve inches thick, and the bed of small stones was very inefficiently applied on many places, and at some points all but omitted. My examinations made me aware of these shortcomings, though not of their extent, but as the time necessary for a complete correction of the work then would proportionally delay the first delivery of water to the city, and my inexperience in this kind of paving interfered with my usual caution, it was, with some exceptions, allowed to pass.

The water began to flow into the reservoir in November, 1858, and continued to rise in it slowly through March and April. During the month of April some very severe gales occurred when the reservoir was a little over half full of water. The position is a very exposed one, and on the south side the waves ran four to six feet up the slopes, the water left within the paving as the waves fell, following behind with some velocity and carrying each time a portion of earth with it. This left pieces of the paving without support and it settled in spots three to six inches. Whether the paving would have suffered, had the work been kept up to the specified requirements, cannot now be known. A certain amount of settlement of the paving must follow the first filling of any reservoir of this kind; the work is then rammed home, and if of the proper character, remains afterwards intact, of which we have no want of examples. I will not pursue this matter further, although it was much talked of at the time, and I allude to it at all mainly in the interest of engineers.

While the gales referred to lasted, the paving was merely watched and driven home as it settled, and well pinned up with chips. After the stormy weather had subsided, we proceeded to repair what was amiss, and to render the whole work of paving secure against any similar contingency, in the following manner.

One of the compartments of the reservoir having been emptied, gangs of men were set to work to drive home the paving, to bring up any pieces, that had settled irregularly, to the correct lines, to remove all the small stones and chips of the pinning, and with small iron tools, made for the purpose, to withdraw any earth lying between the stones. This having been done, a gang of masons followed these, and with well-prepared cement-mortar filled up all the joints and openings, working it in with their trowels, and pinning it at the same time with the chips and spauls which had been withdrawn by the first gang. This operation was much less expensive than had the paving been taken up and relaid in mortar, and the result has been very satisfactory. The workmen were directed to leave open joints occasionally, that in the case of the reservoir water falling rapidly any water lodged behind the paving-stones might be able to follow it.

After our experience of the action of the wind on this reservoir, and my examinations since of dry stone paving upon other reservoirs, both here and in Europe, I should make the thickness of dry paving not less than eighteen inches, and lay it with stones capable of close work, upon any reservoir as much exposed to the action of wind as is the Ridgewood reservoir.

The influent chamber is placed at the south end of the division embankment, and the effluent chamber at the north end of the same embankment. Each chamber is arranged to communicate with both compartments of the reservoir, or with either compartment, at will. The distance between the two chambers is 1,215 feet. The water, therefore, received into either compartment of the reservoir, from the influent chamber, has more than this distance to travel (see plate 38) before reaching the chamber whence it is delivered to the city, and during its imperceptible progress, as regards velocity, between these two points, it has ample time to deposit any sediment with which it may have been charged. But as it very rarely happens that the water in the conduit or pump-well is affected in this way, it usually reaches the reservoir clear as spring water. For this character of water, two divisions in the reservoir would not have been necessary, as providing from one to the other for the intermittent retention necessary to subsidence, one compartment would have equally satisfied the necessities of the case as to that point; but the two are necessary as a means of cleaning and repairing the reservoir without drawing off more than half of its reserve of water.

The influent chamber is in length, twenty-eight feet, width, nineteen feet; the bottom is situated six feet below high water of the reservoir, and four and a half feet below the centre of the mouth of the delivering pipes. From this pool, or chamber of water, an open passage communicates with the western division of the reservoir, and another with the eastern division. Either of these passages can be shut off by flash-boards, and the whole delivery, in that case, thrown into the opposite division. The water, flowing through these passages, falls, when the reservoir is low, into a shallow well of water, placed there to protect the paving of the slope from the wear of the falling water; thence it reaches the reservoir over a brick paving set on edge, laid in mortar, and resting on the heavy stone-work of the foundations. A portion of the bottom of the reservoir is paved here, to defend the bottom when the water first touches it. This paving is of stone, laid in hydraulic mortar. These last details are not seen when the reservoir is full.

The masonry of the work consists of granite, carried up in courses, the face-stones being cut in bed and build, and dressed to the lines of the work. The whole is laid in hydraulic mortar, composed as already described. The drawings will give the details of the foundations, &c. (See plate 39.)

The influent chamber is large enough to receive the terminal pipes of four force mains, being the number necessary to deliver the waters of four engines, each of ten millions gallons daily capacity, covering, therefore, the forty millions of supply contemplated in the design of these works, half of which supply is provided for now, as previously mentioned.

The chamber shows but two delivery pipes now, being the mouths of the force mains in current use. These terminal pipes are carefully built into the masonry, the back of which, in contact with the earthen embankments, is carefully puddled all round, this puddle being connected with the puddle of the reservoir. A separate piece of masonry, situated at the foot of the exterior slope of the bank, holds and envelops the mains there also, and secures the pipes from any longitudinal motion within the reservoir grounds, and from the leakage which such motion might entail. An inspection of the sharp inclination upon which the force main pipes are laid, below the reservoir, will show the risk of some such effect being produced there.

The effluent chamber (see plate 40) is arranged so as to connect the city supply mains with the water of either division of the reservoir, or with both, at convenience. My object was, in both chambers, to simplify, as much as possible, the connection of the mains with the reservoir compartments, and at the same time to make their pipes easily accessible for repairs, complicating as little as possible, under such circumstances, the reservoir works.

The water space of the effluent chamber is connected by passages eleven feet wide, with the two divisions of the reservoir. A heavy granite wall is built across each passage, rising to the same level as the top of the reservoir banks. In each wall there are four openings, the two lower openings being 3x3 each, and the two upper openings 3x4 each. Iron sluices running in iron slides, faced with composition metal, cover and control these openings. From these sluices, iron rods of two inches diameter rise to the top of the work, where they terminate in screws and gearing for the movement of these sluices. The faces of these iron sluices are parallel;—it is evident now that they would have been tighter, had the sluices been wedge-shaped, like the sluice-gates of ordinary stop-cocks. The possibility of their getting fixed in that case, induced us to have them made as they are.

In front of the sluices, towards the reservoir, in each passage, copper wire screens are placed, twenty-two feet in height, to prevent fish, leaves, &c., from passing into the effluent chamber, and so into the supply pipes. As a further precaution, a screen of similar material defends the pipe mouth.

Immediately behind the effluent chamber proper, but connected with it, there is a dry chamber, open to the surface, except as it is now covered by a moveable iron roofing. The supply mains pass through this dry chamber, and it is here that the stop-cocks of these mains, and the stop-cocks of the waste pipes are placed. Into the granite wall, six feet thick, separating this chamber from the water chamber, the three mouth-pipes of the three pipe-mains, each of thirty-six inches diameter, are built in place. There is but one of these mains in use now, and but one large stop-cock in the chamber at present; the mouths of the other pipe-mains are for the present closed in front. Into the opposite wall of the stop-cock chamber, pieces of the same sized mains are built, in order that when a second or third main is required to be laid, it may not be necessary to break into any of the masonry. In the same chamber the stop-cocks of the waste pipes are found. These waste pipes are of twelve inches diameter, and communicate with each division of the reservoir, their stop-cocks being closed, except when, in the course of drawing the water off either division, the bottom is desired to be drained off thoroughly. This drainage water is carried by a twelve-inch pipe to a pond hole on the opposite side of the turnpike road. The mouths of these drain pipes are outside of the copper screens, as will be seen by the drawings.

The bottom of this chamber as well as of the effluent chamber proper, is paved with hard-burnt brick, set on edge, and laid in cement mortar. The masonry is of blue stone, finished with coursed granite, except the heavy foundations, which are of rubble work. The whole work is laid in hydraulic cement mortar, of the character elsewhere described. The earthwork of the division embankments, where it connects with the masonry was carefully rammed, and the puddle wall of the embankment was widened there, so as to cover the whole space between the buttresses. The puddle was enlarged in the same manner behind the walls of the influent chamber.

The apparatus for moving the sluices is protected by a small house built over each passage.

The paving of the reservoir slopes, where they meet the top lines of the banks is finished by a dwarf wall, and blue stone coping, upon which there is placed a low iron fence to keep visitors and children off the water slopes.

To the west of the entrance gate of the reservoir a neat house is built for the keeper of the grounds. The reader is referred to the designs of this house and plan (plate 41). It is neatly constructed, of the materials and workmanship usual in this class of house, which, as having no special relation to water works, it is not thought necessary to describe.

Within the engine-house grounds it should have been mentioned that a similar house has been built for the engineer of the pumping engines, of less architectural pretensions than the one at the reservoir, but equally serviceable and well built. (See plate 37.) A keeper's house also exists within the Jamaica pond grounds, but this house was there when the grounds were acquired for the works; it has since been repaired and made serviceable for the family of the person in charge at that station.

## PIPE DISTRIBUTION.

The contract provided for the laying of the following lengths and sizes of pipes:

| | |
|---|---|
| 36-inch mains | 5 miles. |
| 30 " " | 5 " |
| 20 " " | 4 " |
| 12 " " | 12 " |
| 8 " " | 30 " |
| 6 " " | 64 " |

The hydrants not to exceed eight hundred. Four inch pipes were used to connect the hydrants with the street pipes; the aggregate length of these for eight hundred hydrants depended upon the widths of the streets, and other considerations, and could not, therefore, be defined.

The stop-cocks laid with these pipes, under the contract, were in number as follows:

| | |
|---|---|
| 36-inch stop-cocks | 3 in number |
| 30 " " | 9 " |
| 20 " " | 17 " |
| 12 " " | 65 " |

| | | |
|---|---|---|
| 8-inch stop-cocks | 168 | in number. |
| 6 " " | 402 | " |
| with | 6 | blow-offs |
| and | 50 | air-cocks. |

I shall confine the word "main" to the pipes of twenty inches diameter, and upwards. These large pipes, or "mains," supply the pipes of twelve inches diameter and under, and these last supply the water takers, by means of "service-pipes" so called.

The mains are never allowed to be tapped, a process which is capable of splitting a pipe, a contingency that very rarely occurs, but to which a large pipe would be more exposed than a small one, and in that case, besides occasioning much inconvenience, it might produce considerable damage.

At this date the amount of pipes, stop-cocks, &c., laid, is as follows:

| | Miles. | Stop-cocks. | Blow-offs. | Air-cocks. |
|---|---|---|---|---|
| Of 36-inch diameter | 5 | 4 | ...... | ...... |
| " 30 " " | 4.771 | 8 | ...... | ...... |
| " 20 " " | 5.929 | 20 | ...... | ...... |
| " 12 " " | 18.033 | 105 | ...... | ...... |
| " 8 " " | 38.935 | 231 | ...... | ...... |
| " 6 " " | 98.725 | 638 | ...... | ...... |
| " 4 " " | 0.348 | 8 | ...... | ...... |
| | 171.741 | 1,014 | 15 | 50 |

With 1,291 hydrants.

There has been no increase in the mains since the contract was completed, but the increase has been great in the small pipes. It will be observed that the extent of thirty-inch pipe laid is less than the contract allowance, while the extent of twenty-inch pipe laid is more.

It was found desirable to connect the centres of Williamsburgh and Brooklyn by a twenty-inch main, and to meet the increased length of twenty-inch pipe which this arrangement rendered necessary, a proportionate reduction was made in the weight of thirty-inch pipe to be supplied; seven hundred and fifty-nine feet of thirty-inch pipe having been exchanged for fourteen hundred and fifty feet of twenty-inch pipe.

At this time the Water Commissioners had no power to do any work except through the contractors (Welles & Co.), nor any power to exceed the expenditure of four millions two hundred thousand dollars prescribed in the contract. The legislature afterwards gave the Commissioners power, with the consent of the Common Council, to make such further provision for laying pipes as they should judge necessary. And it gave them at another time power to have prepared a plan to govern the sewerage of the city and to initiate the construction of the requisite sewers under that plan, to which I will recur hereafter.

In the pipe distribution of Brooklyn the task of the Engineer was directed, as well as limited, by the following clause in the contract specifications: "The character, arrangement, and number of branches, and other appurtenances herein provided for, to be determined by the practice of the Croton Aqueduct Department, and the Cochituate Board at Boston, with such modifications as local circumstances may require."

The weights of the pipes, the number and character of the stopcocks and special castings, were thus made dependent on the practice of these two cities, a practice which was quite as much noted for its differences as for its conformities. These conditions prevented us from going far wrong, but they at the same time opened a large field for debate between the Engineer and the Contractors; the results of these troublesome investigations, however, were generally creditable to the fairness of the Contractors.

In planning the position of the pipes upon the city map, I gave my attention more particularly to the position of the large mains, and so determined them as to accommodate, with as little expenditure of small pipes as practicable, the *then* centres of the population. The positions of the *small* pipes and of the hydrants were left, under general directions, very much to the judgment of the Assistant Engineer in charge of the pipe drawings, who performed this duty zealously and attentively.

The points which required the most consideration in this connection were the weights of the pipes, the number of the stopcocks, the character of the stopcocks and hydrants, and the number of extra branches which should be laid now, especially in the suburbs in view of the future wants of the city. A small alteration in any of these particulars involves a considerable saving or expenditure of money when the materials are iron, lead, and copper.

In the cities of New York, Boston, and Brooklyn, we may take the difference in height between the receiving-reservoir and the mean tide in each case as representing sufficiently the extreme pressure to which any of their pipes can be subjected, not including in this the ram which may be produced by accident or carelessness.

This operating head of water is as follows, in round numbers:

| | |
|---|---|
| For New York | 115 feet. |
| For Boston | 120 " |
| For Brooklyn | 170 " |

The pressure to which a pipe may be subjected in Brooklyn exceeds therefore by fifty-five feet the pressure to which it can be subjected in New York, and by fifty feet the pressure to which it can be subjected in Boston.

To meet this difference in head, a difference in strength or weight of pipe became necessary. To have increased the weight for the whole city would have been unfair towards the Contractors, and unfair towards the city, as initiating a system of pipe weights which in their future operations would be likely to be disregarded as being for the larger half of the city unnecessarily costly. I recommended, therefore, to the Commissioners that two classes of pipes should be adopted. The first class, which was marked "A," to be made applicable to all of the city situated above a plane one hundred and twenty feet below the receiving reservoir. The second class, marked "B," to be made applicable to all the streets of the city lying between this plane and tide-water, and subject, therefore, to a pressure exceeding one hundred and twenty feet of head and not more than one hundred and seventy feet.

The weights of pipes in use in various cities of the United States and of Europe were obtained and collated but with little service as regards any governing rule, the practice of engineers had been so various and their experience in hydraulic works of this kind so comparatively recent.

The weight was found to increase with the head, but generally in a greater proportion than the differences of height or head seemed to justify.

As regards the first class of pipes, the extreme pressure upon which was intentionally made to assimilate closely to the cases of the cities of New York and Boston, the mean weights arrived at were necessarily founded on their practice.

As regards the second class, the following formula was used to assist in determining the additions in weights to be made to the weights of the first class, to meet the greater strain upon the pipes of the second class:

$$t = \frac{5pr}{c-p} + x$$

$t$, being the thickness of the pipe in inches;
$p$, the pressure due to the head of water in pounds per square inch;
$r$, the interior radius of the pipe in inches;
$c$, the allowed cohesion of the iron in pounds;
$x$, a constant added for defective casting, rust or life of the pipe, and handling.

In Hodgkinson's experiments, cast-iron broke under a tensile strain per square inch variously of 13,434 pounds, 14,525, 16,676, and 21,907 pounds, and he recommends that it should not be loaded or strained to more than one third to one fifth of its breaking strength.

Our iron, so far as we have tested it, showed a higher standard of strength than the average of the above figures, and besides this general characteristic of American iron, compared with English, both as regards the ordinary qualities of either cast or wrought iron, our pipe castings cannot meet the proof to which they are subjected before acceptance, unless the iron is of a class decidedly superior to that of common castings.

The iron of our pipe castings, so far as tested, bore a breaking strain of 20,000 to 22,133 pounds per square inch.

In the following table I give the prevailing weights then in use at New York and Boston ; next the weight adopted by us for the "A" class of pipes, and next the results of the above formula, taking the cohesion of iron at 7,500 pounds, equal to one third of our best pipe iron, and again at 5,000 pounds, equal to one fourth of our ordinary pipe iron.

Genieys, in his statement of the weights of the Paris pipes, and the rule by which they are computed, takes $x$ at 0.39 inch. In the first application of the above formula I have taken it at 0.40. In the second application of the formula, where $c$ is made equal to 5,000 pounds, $x$ is taken as follows:

For 36-inch pipes, $x$ = 0.24 inch.
" 30 " " = 0.25 "
" 20 " " = 0.28 "
" 12 " " = 0.32 "
" 8 " " = 0.33 "
" 6 " " = 0.34 "
" 4 " " = 0.35 "

Five times the actual pressure ($5p$) is applied in both cases as no more than a proper allowance, for the beat or ram, caused by breaks in the pipes, or the too sudden shutting off of the water by stopcocks or hydrants. This is the allowance made by Genieys. In all other respects the formula is Barlow's.

NINE FEET PIPES.

| Diameter of Pipe. | Prevailing Weights at | | Adopted weights. Brooklyn. | Weights by Formula for 120 feet head, Cohesion taken at | |
|---|---|---|---|---|---|
| | Boston. Head 120 feet. | New York. Head 115 feet. | Head 120 feet. Class A. | 7,500lbs. | 5,000lbs. |
| | lbs. | lbs. | lbs. | lbs. | lbs. |
| 4-inch | 220 | 180 | 200 | 196 | 189 |
| 6-inch | 335 | 336 | 330 | 307 | 303 |
| 8-inch | .... | .... | 430 | 438 | 438 |
| 12-inch | 700 | 672 | 680 | 730 | 760 |
| 20-inch | 1,700 | 1,551 | 1,600 | 1,470 | 1,580 |
| 30-inch | 3,000 | 3,110 | 3,000 | 2,660 | 2,984 |
| 36-inch | 3,600 | 3,745 | 3,600 | 3,510 | 4,045 |

The discrepancies between either application of the formula and the common practice are here well marked. Both applications of the rule show the prevailing weights for the six-inch pipe to be too great, while those for the twelve-inch pipes are rather light. The second application with the cohesion, taken at 5,000 pounds, gives, on the whole, the safest results, and results corresponding as near as could be expected to a practice founded mainly on the chance judgment of practical men, unlimited, apparently, by any strict reference to the relative strains of the different sizes of pipes. This application requires the thirty-six-inch pipe to weigh 4,045 pounds for one hundred and twenty feet of head, a weight, however, which we could not adopt, the contract confining us in this respect to the practice of the cities of New York and Boston.

Having determined by that practice the weights as above, for the "A" class of Brooklyn pipes, I used this formula (the results of which are given in the above instance, merely to enable engineers to judge of its reliability for the purpose in view) to obtain the additions which should be made to these, to get reliable weights for the "B" class of pipes, exposed to a head of one hundred and seventy feet, as well as to enable me to show to others that the additions recommended by me were reasonable.

I will here repeat the weights of the "A" class of the Brooklyn pipes, and place opposite these the additions required by the formula, or the differences between the formula weights for 120 feet head, and for 170 feet head; and in the fifth column, will show the weights adopted for 170 feet of head.

BROOKLYN, NINE-FEET PIPES.

| DIAMETER OF PIPE. | Weights of Class A, 120 feet head. | Additions required, as per Formula, to make class B. | | Class B pipes, 170 feet head, Weights adopted. |
|---|---|---|---|---|
| | | Cohesion taken at 7,500lbs. | Cohesion taken at 5,000lbs. | |
| | lbs. | lbs. | lbs. | lbs. |
| 6-inch.................. | 330 | 26 | 41 | 360 |
| 8-inch.................. | 430 | 47 | 72 | 500 |
| 12-inch.................. | 680 | 104 | 158 | 820 |
| 20-inch.................. | 1,600 | 289 | 430 | 1,900 |
| 30-inch.................. | 3,000 | 624 | 957 | 3,700 |
| 36-inch... .............. | 3,600 | 904 | 1,374 | |

Our weights for the "B" class of pipes were much complained of as being unwarrantably heavy; but these calculations show that the complaint was unreasonable. There were no 36-inch pipes required of the "B" class. The 4-inch pipes, which, on the Brooklyn works, are only permitted to be used for hydrant connections, were all made of one uniform weight.

Mr. Neville, in the new edition of his work on hydraulics, published in 1860, gives two expressions whereby to obtain the thickness of pipes; but they neither of them correspond with our practice in the United States. They are:

$$t = .0024\ (n + 10)\ d + .33\text{—for pipes cast horizontally.}$$

$$t = .0016\ (n + 10)\ d + .32\text{—for pipes cast vertically.}$$

Mr. Neville gives also the equivalent of the formula used by M. Dupuis, Engineer of the Paris water works, viz.:

$$t = .0016\ n\, d + .32 + .013\, d.$$

This also gives results much below the weights in use in the United States.

In these expressions, $t$ is the thickness of the pipe in inches;

$n$, the number of atmospheres of pressure, taken at 33 feet each, to which the pipe is to be subjected;

$d$, the diameter of the pipe in inches.

None of these expressions allow sufficiently, in my opinion, for the difference of head. After the pipes are laid in the ground, provided that they are sound when laid, they are seldom broken except by the ram caused by the hasty shutting of stopcocks. We have many instances of pipes being burst from this cause. With 100 feet of head, the pressure within the pipe is at the rate of 43 lbs. to the square inch of surface; with 200 feet of head, it is at the rate of 86 lbs. per square inch. When a wave or ram is brought into action, it bears in its force a positive relation to these pressures. But the above formulæ allow very little additional thickness for this increase of head, presuming, apparently, that the liberal allowances inherent in all pipe-castings, for the contingencies of handling, and rust or life, will meet in large part the other case. I do not think it right to stretch these allowances to other than their proper application—they are none too much now for the duties required of them. The following modification of M. Depuis' formula, more nearly meets our practice, and allows better for the effect of difference of head. It gives results, however, somewhat below our usual weights, but results that under the late improvements in castings, may prove sufficient:

$$t = 3.4\, n\, (.0016\, d) + c,$$

$c$ being the constant, which, in this expression, must vary with the diameter of the pipe, as follows:

| | |
|---|---|
| For 6-inch pipe...................... | $c = 0.40$ inch. |
| For 8-inch pipe...................... | $c = 0.40$ " |
| For 12-inch pipe...................... | $c = 0.39$ " |
| For 20-inch pipe...................... | $c = 0.38$ " |
| For 30-inch pipe...................... | $c = 0.37$ " |
| For 36-inch pipe...................... | $c = 0.36$ " |

The aggregate lengths have already been given of the pipes to be laid under this contract. These were all cast in 1857 and 1858, and were all laid in the years 1857 to 1859.

The pipes were cast at six different foundries—

| | |
|---|---|
| The Conshohocken foundry, Pa............ | S. Colwell & Co. |
| The Florence foundry, N. J.............. | Jones & Co. |

The Warren foundry, Phillipsburgh, N. J.
The Lumberton foundry, N. J............ Irick, Cole & Co.
The Millville foundry, N. J.............. R. D. Wood & Co.
The Phœnix foundry, Glasgow, Scotland... Thos. Edington & Sons.

Before these pipes could be accepted, they had, as usual, to be thoroughly cleansed and proved. The mode of proof is thus stated in the words of our pipe specifications:

"Every pipe, branch, or casting, of whatever form, shall pass a careful hammer inspection, under the direction of the engineer, or his inspector, and shall be subject thereafter to a proof by water pressure of three hundred pounds to the square inch, for all pipes of thirty inches diameter, and under, and of two hundred and fifty pounds per square inch for all pipes exceeding thirty inches diameter. Each pipe, while under the required pressure, shall be rapped with a hand hammer from end to end, to discover whether any defects have been overlooked."

To explain more fully our practice, and to abbreviate as much as possible what might otherwise be said here, the pipe specifications are given in full in the Appendix.

Although that practice was not different, except in the water proof, from the practice prevalent with English engineers, whose experience in these matters was, at that time, more extensive than ours, nor at all different in the water proof from the practice in the United States, of our largest cities, and best engineers, yet the faithful application of it by our inspectors met with much complaint and objection from every foundry, and we were accused, as is usual in such cases, of unnecessary and impracticable nicety in our tests and examinations. A persistence, however, in our determination to obtain a sound and reliable character of pipes, and to discard and search for all defective ones, led gradually to a change and improvement in the character of these castings, brought about partly by the use of a better mixture of metals in the cupola, and partly by more careful manipulation in the moulding.

Since the commencement of our work in 1857, the character of all pipe castings has, from this and other causes, been in some respects much changed. At that date the practice in the United States was to cast all pipes below twenty inches diameter, and at some foundries, including those of twenty inches diameter, horizontally, or with the moulds lying at a slight inclination. The effect of this mode of casting was to produce pipes of a variable thickness of iron. A pipe of uniform thickness throughout was very rarely obtained by this mode, though the irregularity in this respect was rarely sufficient to compromise the ability of the pipe to meet the proof. But the knowledge of the fact inclined engineers to make the pipes so much heavier; the thin portions, though sufficient for present strength, left, in that case, but a small margin for what we call the life of the pipe. It has been already seen that in addition to the thickness of iron simply necessary to meet the daily working strain, a considerable addition is always intended to be made to meet the contingencies of handling and durability. This important addition may be, and often is, lost in effect in the pipes cast horizontally. In pipes cast vertically its presence can always be depended on. It is not the interest of foundries to vary their processes, such changes always leading to present outlays, which business men naturally desire to avoid. It is, therefore, to the credit of the managers of the Warren Foundry that they promptly sympathized with my desire to remedy, where practicable, the defects in uniformity of thickness of our pipe castings, and were the first in our country to cast pipes vertically of diameters below twenty inches, ultimately applying this mode successfully to six-inch pipes.

General Meigs, when in charge of the Washington Water Works, as well as the other engineers of the country connected with such hydraulic works, lent their aid and countenance heartily in the same direction, and it may be said now, I believe, of all our foundries, that they have adopted and approved the improvement indicated, and have all but discarded the old mode of casting pipes, except for those of eight-inch diameter and under.

At the Phœnix Foundry, in Scotland, pipes of all diameters were, at this time, (1857) cast vertically, the larger diameters in twelve-foot lengths; but this was the only foundry in Great Britain, if I mistake not, that at that time extended this practice to all sizes of pipes, a practice initiated at this foundry, and connected there with a patent for the rapid preparation of their moulds and cores. These pipes were at that time used by Mr. Bateman, the Engineer of the Glasgow Water Works.

The inspection and proving of the pipes was done at the several foundries. An inspector was stationed for this purpose at each foundry, and each foundry procured or supplied the necessary proving presses, to meet the requirements of the specifications. The Ashcroft gauge, the Allen gauge, and a mercurial gauge, were variously connected with the presses to enable the inspector to ascertain that the requisite pressure was applied.

In English practice, the hydraulic proof is usually limited to twice the head of water that the pipe may be subjected to. In the United States this proof is extended to at least four times the extreme head. If we consider the ram of water to which the pipe is subjected by the careless shutting of stopcocks, and by the rapid shutting of hydrants at fires, our proof will not be found to be in excess of the duty required of the pipe. In testing the pipes, however, the free use of the hand-hammer is the surest mode of discovering defects. The water proof tries the general strength of the iron, and will often expose defects which the first hammering of the pipe did not happen to touch. It has been especially serviceable sometimes in detecting mixtures of inferior iron used in the cupola, without the knowledge of the inspector.

Mr. John H. Rhodes, who acted as inspector of the pipe castings at the Conshohocken, and afterwards at the Phillipsburgh foundry, besides his current duties, interested himself specially in noting the nature of the defects usually prevalent in pipe castings, and the extent of variation in thickness, brought about by the horizontal mode of casting. In this respect he rendered valuable service.

In boring the pipes after they are laid (technically called tapping), for the purpose of conveying the water by small service-pipes to the adjoining houses, a certain thickness of iron is necessary in the pipe to hold the brass connection, which is called the tap. To insure a sufficient thickness, belts are cast on the street-pipes of twelve inches diameter, and under. These belts increase the thickness of the pipe where they occur about three eighths of an inch. The pipes are tapped only at these belts; were they allowed to be tapped between the belts, the iron would frequently be found too thin to hold the tap. In many of the pipes cast horizontally, it does not exceed one fourth of an inch in thickness on the thin side of the pipe, being proportionally too thick, in that case on the opposite side. This defect has always been known of that class of pipes, and hence the belts.

In the process of casting pipes vertically, these belts cannot well be introduced, and hence, for all pipes cast that way, they are dispensed with. At first view this was considered objectionable; but it is found in practice that when the weight of iron allowed the pipe is uniformly distributed, it gives a sufficient thickness for all the purposes of tapping.

The introduction of pipes of twelve feet in length instead of nine feet, reduced the cost of laying by reducing the number of joints in a mile of pipes, and to that extent reducing the amount of lead, yarn, and labor, required. The first twelve-feet pipes used in this country were those sent from the Phœnix Foundry, Glasgow. In their contract with that foundry for pipes, H. S. Welles & Co. ordered some 12-feet lengths of thirty-six inches diameter as an experiment. The Warren Foundry very soon took up this improvement and extended it successfully to pipes of twelve and eight inches diameter. They can now be obtained of this length at any of the prominent pipe foundries of the country.

The letter "A" was cast upon all pipes of the class "A," and upon all the special castings applicable to this class of pipes. The letter "B" was cast upon all pipes and castings of the class "B." Each pipe was beside numbered, the number being cast upon the pipe. It sometimes happened that pipes which had been rejected at the foundry by the inspectors, got, notwithstanding, accidentally forwarded to Brooklyn. The inspector's return enabled us, by the numbering, to detect these with little difficulty.

From a number of observations made by the Assistant Engineer, the lead used at the joints in the laying of the pipes was found to average as follows. This can only be considered a near approximation, as no record was kept of the aggregate amounts of lead used by the contractors on each size of pipe.

As a matter of curiosity I have added the averages of lead used in English practice, in the cities of Edinburgh and Glasgow in 1857, as furnished me by the Engineers of those cities, Mr. J. Leslie and Mr. D. Mackain, to whom I am otherwise indebted for valuable information in this connection, very obligingly communicated. The experience is also added of that part of London then under the superintendence, as regards water, of Mr. Simpson.

AVERAGES OF LEAD USED AT THE PIPE JOINTS.

| DIAMETER OF PIPE. | LEAD USED PER JOINT, IN POUNDS. | | | |
|---|---|---|---|---|
| | BROOKLYN. | EDINBURGH. | GLASGOW. | LONDON. |
| 4-inch | ...... | 6 | 5 | 4 |
| 6-inch | 8.50 | 7.50 | 7.50 | 6.50 |
| 8-inch | 10 | 10.50 | 10 | |
| 12-inch | 16 | 15.25 | 17 | 16 |
| 20-inch | 30.30 | 31.50 | 35 | ...... |
| 30-inch | 53.50 | 50 | ...... | 43 |
| 36-inch | 66 to 73 | 60 | 60 | ...... |
| 36-inch force main | 170 | ...... | ...... | ...... |

Some cast-iron pipes have been used since 1857 in Great Britain, dispensing with the use of lead calking for diameters of twelve inches and under. The spigots and sockets of the pipes are in this case turned and bored for about half the lengths of the sockets so as to fit into each other snugly (plate 45). In the process of laying, Roman cement is used instead of lead, laid as red-lead is laid on the turned portion of the joint, the spaces left on the remainder of the joint being filled with the same material. Some of our gas companies have lately used pipes of this character, but I am not aware that they have yet been applied in the United States to water service.

A portion of the pipes received for our works from Glasgow, Scotland, were coated with a preparation of coal-pitch and oil, according to Dr. Angus Smith's patent process. This protects the pipe from rust before it is laid, and has thus far proved an excellent protection from rust after being laid. The durability and success of this process has been tested fourteen years in England. In the United States it has not been anywhere in use more than six years, the first pipes prepared in this way having been used in 1858 upon our works. Since that time various of our foundries in New Jersey and Pennsylvania have provided the proper apparatus to apply Dr. Smith's protective process to the pipes cast by them, and the cities of New York, Boston, and Philadelphia, besides other places, have lately had all or a portion of the pipes cast for their current requirements protected in this way. The extra cost of the application on the Scotch pipes was one dollar twenty-five cents per ton. In the United States it has cost from one dollar twenty-five cents to one dollar fifty cents per ton. Some

6

of the first pipes (thirty and thirty-six-inch) protected for the Brooklyn works in this way may now be seen at Williamsburgh, and some lying within the Ridgewood reservoir grounds; they consist of pipes held in reserve for accidents or repairs, and have been lying out of doors since 1857. This kind of exposure is a very severe test of any covering of the kind.

In England Dr. Smith's patent secures him a remuneration for his application, but in the United States, where he has no patent, I have not been able to obtain for him any compensation for the use of his process, although it has been of great value to us, both as a defence of our iron pipes from rust and as a defence of the water from the discoloration which rust at times produces.

At the foundries the application of the process, as it involved outlay, met at first with opposition, and therefore any promise of remuneration from the foundry companies to the inventor was out of the question; in the case of our city authorities, a claim of this kind would have engendered serious opposition to the use of this valuable process at all. I say thus much apologetically to Dr. Smith, who, notwithstanding, furnished me, unconditionally, with such information and directions in regard to the process as, in conjunction with the information received through our inspector at Glasgow, has enabled me to furnish our foundries with the necessary directions for its successful application in the United States. The condition of its application to pipes will be found attached to our pipe specifications as given in the Appendix.

The tubercular corrosion so common heretofore in cast-iron water pipes has been made a serious objection to their use as conductors of soft water for city purposes; with hard water, it is well understood that this kind of corrosion does not take place. The chief objection urged has been in reality the least important in practice. I refer to the asserted vitiating effect of the tubercles on the purity of the water, and incidentally to its discoloration. The first effect was never exhibited except at dead points, viz., at the termini of pipes or at points where, from whatever reason, the water remains for a considerable time motionless; at such points the water will after a time become stagnant and disagreeable, whatever be the nature of the pipes employed. The discoloration could only occur when, as in the case of fires or accidents, considerable velocity is engendered in the pipes. The ordinary velocity of the water does not produce a sufficient current to remove the tubercular matter. This objection, although of rare occurrence, was yet of a nature to produce dissatisfaction among water-takers, and to get rid of it altogether was very desirable.

The most serious evil produced by the tubercular corrosion was the consequent reduction in the interior diameter of the pipe and in its delivery of water. In small pipes, of four inches diameter and under, the delivery would sometimes be reduced from this cause more than one half, and in large pipes the effect, though less, was always serious. It involved, in effect, unless corrected, the use of larger pipes throughout the city than would otherwise have been necessary to meet the conditions of the service, increasing seriously in this way the cost of the works.

The life of the pipe, or its durability, was not so much affected by tubercular corrosion, as has been asserted; the tubercles are evidently not composed entirely of what we call rust. A point of rust forms the nucleus or origin of the tubercle, but the matter of which it is composed seems, in large part, to be derived from the fine and almost imperceptible sediment floating in all river waters, whose particles are attracted upon the tubercular centres by some action which we cannot explain.

Early in 1857 the attention of the Water Commissioners was called by me to this evil, with a view to its remedy; and I was authorized to inquire, by advertising and otherwise, whether a preventive could be found and made available for our pipe distribution. The subject of the defence of iron from rust had for many years engaged the attention of both English and French engineers, more particularly in connection with its effect upon iron ships. The British Association for the advancement of Science had instituted experiments to this end, and two reports had been made by Mr. Robert Mallett, the chairman of one of its committees. The result of our inquiries was the receipt of a number of communications, offering, in various ways (most of them untried) to defend our pipes from corrosion.

Of all these proposals, there were but two which could be entertained as being known, and as having been well tried by years of successful practice. The first one of these had been used in France, and was communicated to us by Mons. Le Beuffe, civil engineer of Vesoul. The second was Dr. Smith's patent process, already adverted to.

The French process consisted of a mixture of linseed oil and beeswax, applied at a high temperature, the pipe being heated and dipped into the hot mixture. In all these applications, the thorough cleaning of the pipe, and its entire freedom from rust, are important pre-requisites. Linseed oil had been used successfully in France applied to the pipes under pressure by means of a hydraulic press, without the admixture of wax. In adding the beeswax, Mons. Le Beuffe dispensed with the hydraulic press, thus simplifying the process.

I have little doubt but that this preparation, applied in this way, would have proved a successful protection to cast-iron pipes, and place entire faith in Mons. Le Beuffe's account of his successful use of it in French practice. The only reasons I have to give for recommending Dr. Smith's process in preference, are, that I could refer to a number of cities in England where it had been used, and was being used when we had the subject under consideration, and that I could instance in support of its use the experience of two engineers in England, of high standing in hydraulic works, viz., Mr. T. F. Bateman and Mr. Robert Rawlinson. Our experience does not enable us to say how long this application will remain intact as a defence against corrosion. It will doubtless lose its virtue in this respect after the lapse of years, and the time will depend partly on the character of the water, but more on the manner of the application of this coal-tar varnish. Of the pipes prepared in this way for Brooklyn, and laid in our streets, only a very few cases have occurred where their condition could be examined. In these instances the pipes have been found in a very clean and satisfactory condition.

The forms and dimensions of the sockets of the different sizes of pipes will be found delineated on plates 43, 44, 45, and 46, as well as specimens of the special castings. Specimens of the hydrants and stopcocks, &c., are shown on plate 48. To many of the large stopcocks indices were attached, to enable the operator to understand the position of the valve under all circumstances.

In the return already given of pipes laid by H. S. Welles & Co., there are included some pipes manufactured and laid by the Patent Cement Pipe Co., of New Jersey. The Commissioners authorized the laying of these pipes on a portion of the Williamsburgh, or eastern district of the city, as an experiment. This class of pipe is as well known now in the United States as the cast-iron pipe, and it is by some preferred to it. My long indisposition has not permitted me to watch the operation of these pipes as regards their alleged advantages or disadvantages, in comparison with cast-iron pipes, and I refer to the authorities in immediate charge of the works since their completion, for this kind of information. These cement pipes, however, have been largely used in our smaller cities, and, so far as I know, have been received there with sufficient satisfaction.

The lengths and sizes of cement pipes laid are as follows:

| | |
|---|---|
| Of 20-inch diameter | 1,143 feet. |
| Of 12-inch diameter | 1,573 feet. |
| Of 8-inch diameter | 993 feet. |
| Of 6-inch diameter | 9,182 feet. |

The pipe-yard is situated in Portland avenue. The offices and the wall fronting the avenue are built of brick; the rest of the yard is enclosed by a high wooden fence. The sheds for the pipe presses and special castings are of timber. There are two proving-presses, furnished by Welles & Co., one for small pipes and one for large pipes. These presses were purchased by the contractors in 1858, and were copies of presses then in use. Like all the presses, however, which were employed then at the different foundries where the Brooklyn pipes were cast, it has been found necessary to alter and strengthen them to admit of the actual application of the pressure indicated as the test in our specifications. In this respect, I followed the standard of water proof which had prevailed for many years before our works were commenced; but the lever gauge which was then used in connection with the process, was, we have reason to believe, very unreliable in its indications.

## MOUNT PROSPECT RESERVOIR.

Of that part of the city of Brooklyn which lies to the south of Atlantic street, a certain portion is situated above the level of the Ridgewood reservoir, and cannot therefore be supplied except by a supplementary and separate arrangement. This highest part of the city was but little occupied in 1856, when our works were commenced. To meet its future occupation and wants, the contract required a reservoir to be built on the highest point of this ground, called Mount Prospect. This reservoir has been named the Mount Prospect reservoir.

It is supplied with water by a pumping engine situated on Underhill avenue, whose pumps derive their water from a branch main connected with the thirty-six inch main from Ridgewood reservoir, laid in De Kalb avenue.

The contract requires this reservoir to have a capacity when full of not less than twenty million New York gallons. As built, it holds when full, 20,036,558 gallons, equal to 2,564,679 cubic feet.

Its high-water stands one hundred and ninety-eight feet above tide, or is twenty-eight feet above the high-water of Ridgewood reservoir when full; the depth of water is twenty feet.

The reservoir grounds cover 11.08 acres; the reservoir works occupy about three quarters of this ground.

The shape of the reservoir and its details are given on plate 51.

The embankments are twenty (20) feet wide at top, the slopes inside and outside being at the rate of one and a half horizontal to one perpendicular.

The banks were carried up in layers as in the case of the other reservoir, and each layer was thoroughly rolled with heavy rollers. The puddling material was prepared as already described, the clay in this case being obtained in part from New Jersey. The bottom was puddled two feet in thickness. In the case of the embankments the puddle was laid on the face of the interior slopes; it was two feet in thickness. Upon the face of the puddle when finished and dressed to its lines, three inches of concrete were laid, formed from clean gravel and hydraulic mortar. Upon this preparation a brick slope wall was laid eight inches in thickness, in cement mortar. The top of this brick slope wall is finished with a coping of blue flagging, upon which a low iron fence has been placed. Upon the surface of the bottom puddle, finished off so as to have a fall of six inches towards the drain-pipe, a paving of brick on edge was laid dry, and afterwards grouted with cement grout, in terms of the contract requirements.

The outer slopes of the embankments were covered with soil and neatly sodded. The top surfaces of the embankments, which are four feet above high-water of the reservoir, are gravelled for part of their widths, the remainder being sodded.

On the north side of the reservoir a small influent chamber is built, ten feet by six feet in plan, and 12.5 feet in depth. (See plate 52.) It is arranged to receive the terminal pipes of two force-mains. The existing force-main of twenty inches diameter enters this chamber, at six inches above the paving. The water from the main rises in the chamber and passes thence into the reservoir by a 30-inch pipe, situated at its high-water line. A 12-inch pipe placed just above the level of the

30-inch pipe, is connected with a sewer, and relieves the reservoir should the pumping accidentally be carried too far.

The chamber is built of granite laid in cement mortar, the face stones coursed work, every stone being cut in bed and build and neatly tooled off on the face ; the bottom is paved with stone twelve inches thick, laid in courses ; this paving rests upon two feet in depth of masonry as a foundation. The walls are finished off with a neat granite coping, and the chamber is covered with an iron flat roof. All the work is laid in cement mortar of the quality already described.

On the west side of the reservoir there is built a gate-house, so-called, over the effluent pipes which connect the water of this reservoir with the high pipe service of this part of the city.

When the pipes pertinent to this high district are laid and brought into use, the mains will be shut off by stopcocks from the rest of the city, and the smaller pipes will have no connection with the other city pipes. At present the only main laid is connected with the lower pipe distribution, but an intervening stopcock controls the conditions of the connection at will.

At this date (December, 1864), the Mount Prospect pipe service has only begun to come into use as a separate service. The reservoir, however, was always kept nearly full of water as a reserve of about three days' supply in case of any repairs being required on the one main upon which the city depends now for its daily supply of water. A small portion of this water was allowed to flow into the city daily, through the connection already mentioned, just enough to insure the movement which is necessary to purity, as well as to maintain the pumping-engine in constant working order. I will add here that a branch pipe from the force-main is connected with the effluent pipe on Flatbush avenue. By this means, when the reservoir requires to be cleansed out or repaired, the pumping-engine can pump directly into this high service, using, in that case, the influent chamber as a short stand-pipe. The stopcock on this branch is kept always closed except in the contingency adverted to. It cannot be too often repeated, that these stopcocks (of paramount importance in their places) cannot be depended on when required to be used, unless they are tried at least once a month and their fitness for immediate service constantly tested in this way.

The form and size of the house through which the effluent-pipes pass will be understood by reference to the plan and elevations (plates 53 and 54). It is placed, as will be seen, upon the outer slope of the embankment, and is entirely independent of the interior slope.

The building has two floors: the basement floor situated two feet below the bottom of the reservoir, and the upper floor on a level with the top of the reservoir embankments. The pipes from the reservoir rest upon the basement floor, and the stopcocks which control them are placed there.

There are two pipes leading from the reservoir ; one, a thirty-inch main, for delivery, the other, a twelve-inch pipe, for drainage. In the case of this reservoir, these pipes pass through the embankment, supported on brick walls at the joints, and bedded in a body of puddle; where they enter the walls of the effluent-chamber this puddle is carried up the face of that wall to the top of the bank.

The brick supports alluded to serve also as a defence against leakage along the line of the pipe. At the foot of the interior slope of the reservoir a piece of masonry (plate 53) is built to receive the mouth of the thirty-inch pipe and the copper screens which prevent fish from entering it.

The mouth of the twelve-inch drain-pipe is connected inside with the wing of the same piece of masonry; this pipe, after passing through to the effluent or stopcock chamber, is connected outside with one of the street-sewers. The thirty-inch pipe, after entering the stopcock chamber, is divided into two twenty-inch mains. There is but one of these mains laid at present, the branch for the other being capped. The upper chamber of the gate-house is for the present more ornamental than useful, but when the city extends over this quarter some valuable application will doubtless be found for it.

The walls of the basement-story of the building are built of granite and gneiss, laid in cement-mortar. The exterior, where it is not covered with earth, is of coursed granite neatly cut; the floor of the basement chamber in which the pipes are, is flagged with four-inch flagging.

The walls of the upper portion of the building are of brick, the faces being of Croton pressed brick. The front pillars, quoins, and facings, are of Jersey sand-stone. The mortar for this part was made from Thomaston lime slaked, one sixth, hydraulic cement, one sixth, and clean sharp sand, four sixths.

In all respects the materials and workmanship of the building, its flooring, and roofing, &c., inside and out, are good, but it would be irrelevant to the purpose of this description to allude further to their details.

The grounds connected with the Mount Prospect reservoir are neatly finished off; an iron fence encloses the whole, with an entrance gateway and stairs on Flatbush avenue.

## MOUNT PROSPECT ENGINE-HOUSE AND ENGINE.

The reader is referred to plates 49 and 50 for the details of this engine-house.

The engine-room is forty feet by sixty feet, and large enough for two engines; the boiler-room is forty feet by forty feet; the coal-shed is forty-one feet by thirty feet.

The house is neatly built of brick, with the exception of the foundations, which are of granite. The facings are of Jersey sand-stone. The mortars used are of the same character as those already described for the reservoir-house.

In determining the position of this house, or rather of the subsidiary pumping-engine placed there, I sought in vain for the light which the experience of others always throws on engineering difficulties.

This pumping-engine, which is situated in the heart of the city, does not receive its water from Ridgewood reservoir by a separate main, but receives it by a branch-pipe from the main which is at the same time supplying the city with water. The mass of this water is flowing into the city, part of it into Williamsburgh, and, like all other city mains, complicated as they are by branches, it is difficult to ascertain the varying velocities in the main, or, by calculations without experiment, the available head at any one point.

The engine might have been placed so low that under any circumstances the water would reach it, but it was desirable (while making it safe under all the conditions of the service) to place it as high as possible in order that its force-main to the Mount Prospect reservoir should be as short as practicable. The general conditions influencing the height of the position were these :

*First.* The length of main from Ridgewood reservoir to the point on De Kalb avenue at the crossing of Washington avenue, where the branch main to the pumping-engine diverged: 26,062 feet of thirty-six-inch main.

*Second.* The length of branch-main to the pumping-engine, dependent upon the position of the engine-house, as now situated : 4,600 feet.

*Third.* The maximum flow of water which might prevail in the city main while the engine was pumping.

To interfere as little as possible with the head of water in the city, the engine at this date pumps only during the night. But the service will by-and-by require that it should be at work during the day-hours as well. The maximum flow of the city main was therefore necessarily referred to the flow of the day-hours, and this was based on the initial flow prevailing then (November, 1858) in the New York city mains of the same diameter, found by experiment to be equal to 631,935 New York gallons per hour, taken at 632,000. This was the initial flow assumed for the Ridgewood main ; the main divides at a point 15,637 feet from the reservoir, giving off there a thirty-inch branch to supply the Williamsburgh district of the city ; after passing this point the reduced flow into the city proper was taken at 474,000 New York gallons per hour ; the distance for the reduced flow being 10,425 feet. Both these quantities include the amount applicable to the pumping-engine for the secondary service under consideration.

*Fourth.* The flow of water through the engine-house branch-main. The engine was required to be able to pump at the rate of 156,250 New York gallons per hour. With the branch-main, however, it had been found convenient to connect two twelve-inch pipes delivering into the city. The flow per hour was hence taken at 166,666 New York gallons on this main; equal to 5.92 cubic feet per second.

The branch-main is composed of thirty-inch pipe for a part of its length, and of twenty-inch pipe for the remainder.

A diagram of the position of these mains is given in plate 1.

These several data are given in the following table together with the resulting calculations, as printed in my report of November, 1858:

| LOCATION. | Length of Pipe in feet. | Diameter of Pipe in inches. | Rate of flow in New York gallons, per hour. | Rate of flow in cubic feet per second. | Calculated head to produce the given flow, by the formula $h = 0.00046749 \frac{L}{d} (V + 0.397)^2$ |
|---|---|---|---|---|---|
| Reservoir to De Kalb avenue | 15,637 | 36 | 632,000 | 22.471 | 31.160 feet. |
| De Kalb avenue to Washington avenue ........ | 10,425 | 36 | 474,000 | 16.853 | 12.5663 feet. |
| Washington avenue to Pump well .................. | 3,000 | 30 | 166,666 | 5.9259 | 1.4437 feet. |
| | 1,600 | 20 | 166,666 | 5.9259 | 4.3497 feet. |
| | 30,662 | .. | ...... | ...... | 49.5197 feet. |

The above formula is derived from a formula of Prony's, which has been altered in its constants, so as to correspond very nearly with the results of experiments made in July, 1858, on the ordinary deliveries of the connecting pipes of the New York reservoirs, and of the Jersey City reservoirs.

The maximum flow of the New York thirty-six-inch main in 1858, which was taken as my guide here, was even then in excess of what it ought to have been, consistent with the maintenance of a fair head of water in the dwelling-houses and other tenements of the city. The citizens were then, and are now much more incommoded than is at all necessary, and they submit to the inconvenience partly from unwillingness in the authorities to warrant the expenditure in additional mains necessary to correct the evil, and partly from the ignorance of those suffering from it, that it is within the reach of correction. In Brooklyn we shall fast lose the advantage of our position if we pursue the same line of inaction.

I have taken the liberty of saying thus much in this place, because although my calculations in reference to the Mount Prospect engine were predicated on the New York flow in practice in 1858, I am now conscious, and in this respect have the concurrence of Mr. Craven, the engineer of the New York works, that it is not advisable to allow this velocity of flow to be reached, productive as it is of a greater waste of head than the wealth and importance of the place justify. The rate of 632,000 gallons per hour gives a day velocity of 3.17 feet per second within the pipe at its origin, whereas the initial velocity should not, in my opinion, be permitted to exceed two feet per second. In the Brooklyn main, at this date (October, 1864), it has reached a rate of 2.64 feet per second.

The formula used in the table given above is different from the formulas of the books, in that it takes into account the loss of head due to tubercular corrosion within the pipes. With hard water this effect is not produced, except in a very unimportant degree as the experience on the London pipes shows, but with soft water, unless the pipes are protected from corrosion, the flow of water in a given pipe is seriously reduced.

In our case the water is quite soft, and the pipe mains were not protected with the exception (as an experiment then) of a few of the pipes imported from Scotland.

The usual formulæ applicable to the flow of water through pipes are based on experiments on clean pipes. There was then no formula, that I am aware of, that allowed for the evil referred to. To have used the usual formula would have led me astray by giving a loss of head less than would have been experienced in practice, thus placing the position of the engine-house above the point where the flow under the ultimate conditions of the service could reach it.

We knew then, from inspection, that the New York 36-inch mains were heavily tuberculated; and we know now that the Brooklyn pipes, where they are not coated with pitch, are tuberculated in the same way, and from the same cause, the passage through them of soft water.

The corrections to be made on any calculations based on the formulas for clean pipes, could only be ascertained satisfactorily by an experiment. The Croton water works presented a good opportunity for such an experiment, as did the formation of the Jersey City water works. In the cases of both these works, the engineers in charge of them, Mr. A. W. Craven and Mr. George H. Bailey, permitted me to make the necessary use of portions of their works, and rendered, besides, important assistance during the trials.

The conduit of the Croton works, as then existing, delivered the water into what is called the receiving reservoir (See plate 58). From the receiving reservoir, two mains, each of thirty-six inches diameter, carried the water into what is called the distributing reservoir, 11,217 feet distant. From the distributing reservoir, two pipes of thirty-six inches diameter, continued the flow into the city.

At the receiving reservoir there were, besides, two 30-inch pipes connected with the city, but the stopcocks of these pipes were closed during the experiment.

To make the experiment, the conduit gates were closed so that no water passed into the receiving reservoir. No water was allowed to pass out of it except through the two 36-inch pipes aforementioned, and all the water passing through these 36-inch pipes entered the distributing reservoir. The amount of water so passing out was ascertained at the receiving reservoir by the falling of the water there, measured upon proper gauges placed in still water. The head of water expended on this flow was the difference in level between the water of the receiving reservoir and the water of the distributing reservoir, ascertained by simultaneous observations made at the two places connected, and by leveling between their two assumed zeros.

The experiment in the case of the Croton reservoirs was made under the immediate direction of Gen. George S. Greene, then in charge of the works of the new reservoir. The same experiment was carefully repeated in 1859, with like results.

On the Jersey City works there are two reservoirs—the Belleville reservoir and the Bergen Hill reservoir (see plate 58). These reservoirs were connected in 1858 by a 20-inch main 29,715 feet in length. The Belleville reservoir is fed from the Passaic river by a pumping-engine. During the experiment in this case, the pumping-engine did not work, and the water flowing through the pipe was known by calculating the amounts due to the falling of the height of water in the reservoir. The head expended was the difference in the heights of the surface water of the two reservoirs, ascertained by leveling between them, and by simultaneous observations during the trials. This experiment was conducted by my assistant, C. W. Boynton, by whom the calculations were principally made.

The conditions of these experiments will be found stated more at large in the Appendix. In the following two tables, I give the results, and have added for comparison the velocities as well as the required head which the usual formulæ of the books would have given me. It will be seen that by reason of the tuberculation, &c., the head expended to produce the given flow was much in excess of the usual results of calculation, varying in this respect with the diameter of the main. In my calculations to ascertain the head of water lost, which should determine the position of the Mount Prospect engine-house, the formula used was drawn from the results of these experiments. Since these experiments were made, M. Darcy, a French engineer, has published the results of his researches on the flow of water through different characters of pipes, including the case of corrosion and tuberculation. His corrections agree very nearly with the experimental results above mentioned.

## TABULAR STATEMENT IN REGARD TO THE EXPERIMENTS UPON THE NEW YORK MAINS.

*(Pipe Main between the Receiving Reservoir and the Distributing Reservoir.)*

| 1. | 2. | 3. | 4. | 5. | 6. | 7. | 8. | 9. | 10. |
|---|---|---|---|---|---|---|---|---|---|
| Head lost in feet. | How ascertained. | Difference in level between the points in feet. | Delivery, in New York gallons per hour. | Delivery, in cubic feet per second. | Length of Pipe, in feet. | Diameter of Pipe, in feet. | Calculated Head to produce the given Head by the formula $h = 0.00046749 \frac{L}{d} (V + 0.397)^2$ (I.) | Calculated head to produce the given flow by Prony's formula (2). (D.) | Calculated head to produce the given flow by Hawkesley's formula. (A.) |
| 20.215 | By leveling between the Reservoir surfaces.... | 20.215 | 596,352 | 21.2036 | 11,217 | 3 | 20.1670 | 14.899 | 14.833 |

## SIMILAR STATEMENT OF THE MEAN OF EXPERIMENTS UPON THE 20-INCH MAIN OF THE JERSEY CITY WATER WORKS.

| Head lost in feet. | How ascertained. | Actual head in feet. | Delivery, in New York gallons per hour. | Delivery, in cubic feet per second. | Length of Pipe, in feet. | Diameter of Pipe, in feet. | Calculated Head to produce the given flow by the formula $h = 0.00046749 \frac{L}{d} (V + 0.397)^2$ (I.) | Calculated Head to produce the given flow by Prony's Formula (2). (D.) | Calculated Head to produce the given flow by Hawkesley's formula (A.) |
|---|---|---|---|---|---|---|---|---|---|
| 28.1285 | By leveling between the Reservoir surfaces.... | 28.1285 | 88,231.5 | 3.13712 | 29,715 | 1.6667 | 28.064 | 17.921 | 16.056 |

NOTE (*a*). The first of the New York mains has three quarter circle curves of 90 feet radius, which are taken into account in the calculations. The Jersey City main is enlarged for a distance of 128 feet, to 24 inches diameter; allowance is made in the calculation for this enlargement, and the curves in the pipe.

NOTE (*b*). The formula (I) is used only as an exponent of the experimental delivery of the first section of the New York mains, and also of the delivery by experiment of the Jersey City main, to make a convenient application of their general results to a similar case in Brooklyn.

### FORMULAS USED IN TABLE.

(A.) Hawkesley's $h = 0.0004338027 \frac{V^2 [L + 54\, d]}{d}$

(D.) Prony's (2) $h = 0.00040085 \frac{L}{d} [(V + 0.15412)^2 - 0.02375]$.

(I.) Expressing the mean result of the New York and the Jersey City experiments $h = 0.00046749 \frac{L}{d} (V + 0.397)^2$

Where $V =$ velocity in feet per second.
$h =$ the head in feet.
$d =$ diameter of the pipe in feet.
$L =$ length of the pipe in feet.

## TABULAR COMPARISON OF THE NEW YORK AND JERSEY CITY EXPERIMENTS.

| EXPERIMENTS: Where and by Whom Made. | Diameter of the Pipe, in feet. | Length of Pipe, in feet. | Mean Heads under which the discharge occurs. | VALUES OF THE VELOCITY IN FEET PER SECOND. As deduced from the discharge. | By Hawkesley's formula. (A.) | By Blackwell's formula. (B.) | By Prony's formula (1). (C.) | By Prony's formula (2) (D.) | By Eytelwein's formula (E.) | By D'Aubuisson's formula (1). (F.) | By D'Aubuisson's formula (2). (G.) | By Weisbach's formula. (H.) |
|---|---|---|---|---|---|---|---|---|---|---|---|---|
| On the 20-inch main, between the Reservoirs of the Jersey City Water Works. By G. H. Bailey and C. W. Boynton. 128 feet of this main is 24 inches diameter; allowance is made in this calculation for this enlargement, and the curves in the pipe... | 1.6667 | 29,715 | 30.262 | 1.4924 | 1.97399 | 1.97624 | 1.9206 | 1.91068 | 1.97063 | 1.927302 | 1.92923 | 2.03255 |
| | 1.6667 | 29,715 | 29.758 | 1.51337 | 1.9575 | 1.9597 | 1.9039 | 1.89338 | 1.95416 | 1.9105 | 1.9123 | 2.01294 |
| | 1.6667 | 29,715 | 29.317 | 1.44893 | 1.94293 | 1.94514 | 1.8891 | 1.87838 | 1.93963 | 1.8957 | 1.8976 | 1.99688 |
| | 1.6667 | 29,715 | 28.365 | 1.45795 | 1.91112 | 1.9133 | 1.8570 | 1.8453 | 1.90789 | 1.8632 | 1.8650 | 1.95938 |
| | 1.6667 | 29,715 | 27.302 | 1.39107 | 1.8750 | 1.877115 | 1.8204 | 1.8077 | 1.87180 | 1.8263 | 1.8282 | 1.91797 |
| | 1.6667 | 20,715 | 26.325 | 1.381155 | 1.84115 | 1.84323 | 1.7861 | 1.7726 | 1.83801 | 1.7918 | 1.7936 | 1.87943 |
| Jersey City Water-Works mean result ...................... | 1.6667 | 29.715 | 28.1285 | 1.43795 | 1.90031 | 1.9053 | 1.8489 | 1.837 | 1.8999 | 1.855 | 1.857 | 1.9502 |
| New York 36-inch main, between Receiving and Distributing Reservoirs. By G. S. Greene. This pipe has three quarter circle curves of 90 feet radius, which are taken into account in the Calculation.................. | 3.0 | 11,217 | 20.215 | 2.99967 | 3.5016 | 3.5230 | 3.484 | 3.514 | 3.4917 | 3.4891 | 3.506 | 3.848 |

The formula which very closely expresses the results of the New York, and the mean of the Jersey City experiments is $v = 46.2502\sqrt{\frac{h\,d}{L}} - 0.397$, whence $h = 0.00046749\frac{L}{d}(v + 0.397)^2$

### FORMULAS USED IN TABLE.

(A.) Hawkesley's $v = 48.0125\sqrt{\frac{h\,d}{L + 54\,d}}$

See Civil Eng. and Architects' Journal, vol. xviii., p. 99, line 28.

(B.) Blackwell's, $v = 47.913\sqrt{\frac{h\,d}{L}}$. This formula takes into account the curves of the pipe.

See Hughes on Water Works, Formula (33). p. 336.

(C.) Prony's (1), $v = \sqrt{(2354.9375\frac{h\,d}{L} + 0.00665)} - 0.0816.$

See Prony's Recherches Physico-Mathématiques sur la théorie des Eaux Courantes, § 184.

(D.) Prony's (2), $v = \sqrt{(2494.69\frac{h\,d}{L} + 0.02375)} - 0.15412.$

See Prony's Recherches Physico-Mathématiques sur la théorie des Eaux Courantes, § 210.

(E.) Eytelwein's $v = 47.8731\sqrt{\frac{h\,d}{L + 54\,d}}$

See Memoires de l'Academie des Sciences de Berlin, 1814 et 1815, p. 165.

(F.) D'Aubuisson's (1) $v = \sqrt{\frac{h}{0.000417568\frac{L}{d} + 0.015536} + \left(\frac{0.00003767485\frac{L}{d}}{0.000417568\frac{L}{d} + (0.015536)}\right)^2} - \frac{0.00003767485\frac{L}{d}}{0.000417568\frac{L}{d} + 0.015536}.$

See D'Aubuisson's Hydraulics (Bennett's translation), p. 206 top.

(G.) D'Aubuisson's (2), $v = \sqrt{(2394.82\frac{h\,d}{L} + 0.00814)} - 0.090224.$

See Neville's Hyd. Formulæ (109), p. 118, Simplification of (F) by omitting the constant [0.015536], and reducing.

(H.) Weisbach's $h - 0.015538\,v^2 = 0.015538\frac{v^2\,L}{d}\left(0.01439 + \frac{0.0171552}{\sqrt{v}}.\right)$

See Julius Weisbach's Ingenieur und Machinen Mechanik, vol. i. p. 748.

Where $v$ = velocity in feet per second.
$h$ = head in feet.

Where $d$ = diameter of pipe in feet.
$L$ = length of pipe in feet.

I will here give M. Darcy's formula for tuberculated pipes, derived from his experimental investigations on all conditions of pipes, already referred to. I give it as it has been reduced to the units of the other formulas given in these tables :

$$V=\left(\frac{\frac{D h}{L}}{0.00061812+\frac{0.00005176}{D}}\right)^{\frac{1}{2}}$$

This formula has been simplified by Mr. Lane into the following shape, which is identical in value with the above :

$$V=138.996\left(\frac{D^2 h}{11.942\ L D+L}\right)^{\frac{1}{2}}$$

Say,

$$V=139\left(\frac{D^2 h}{11.942\ L D+L}\right)^{\frac{1}{2}}$$

This formula is of more general application than the modified formula given with our tables, which was not sought to be made applicable much beyond the limits of the experiments by which it was determined.

A simpler formula, and sufficiently correct for ordinary purposes, is the following, derived by Mr. Lane from those given above :

$$V=40\left(\frac{D h}{L}\right)^{\frac{1}{2}}$$

There ought, however, to be little occasion for the use of such formulas hereafter, now that we possess a knowledge of more than one mode of defending cast-iron pipes from tubercles and rust.

| | FEET. |
|---|---|
| Recurring to the first table given and the results of its calculations, we have as the head lost at Mount Prospect........................ | 49.5 |
| To this I have added for the effect of the various branches on the main, as gathered from another experiment........................... | 10 |
| | 59.5 |
| The high-water of the Ridgewood reservoir stands 169.7 feet above tide. The reservoir cannot always be kept full, but it was assumed that would rarely sink below 164 feet.............................. | 164 |
| Deduct head lost as above........................................ | 59.5 |
| | 104.5 |
| And we have 104.5 as the height at which water would stand in the well of the engine-house under the extremes of the service. Call the floor of engine-house above the well 15 feet.................. | 15 |
| And we have as, the height of the engine-house floor above tide....... | 119.5 |

The floor of the engine-house, as now built, stands 121.9 above tide.

For the class of engine which has been built no well was requisite, but for the Cornish engine contemplated in the contract, a well would have been a measure of safety, as defending it from the hourly variations of head observable in the routine city service.

Until there exists in the city main the maximum flow of water allowed in the above calculations, we shall not be able to know how nearly they may approximate to the reality. We have the means, however, of judging of the probable result, by using Mr. R. W. Hamilton's observations of the pressure on the supply-main during the 17th and 18th of February, 1863, when the engine was pumping during the day as well as during the night.

There is an Ashcroft gauge in the engine-room connected by a small tube with the supply or induction-main. The gauge stands at eleven and a half feet above the pump bucket at half stroke, and the position of half stroke is one hundred and ten feet above mean tide. Its readings during the day hours, converted by me into feet, and referred to the centre of each pump as above, were as follows :

| FEBRUARY 17, 1863—HOUR. | Pressure in Feet by Gauge at Pump. | Corresponding Height above Tide. |
|---|---|---|
| 6 A. M. | 37 | 147 |
| 7 A. M. | 25 | 135 |
| 8 A. M. | 19½ | 129½ |
| 10 A. M. | 16 | 126 |
| 11 A. M. | 19½ | 129½ |
| 12 A. M. | 23½ | 133½ |
| 2 P. M. | 25 | 135 |
| 4 P. M. | 24 | 134 |
| 6 P. M. | 25 | 135 |

The consumption of water in the city was at its maximum at 10 A. M. on that day, and the variations in its pressure accord with the variations in consumption forenoon and afternoon. The occurrence of a fire in the city would account for the low pressure at 10 A. M.

From a record kept of the amount of water consumed each month, I find that during the month of February, 1863, the average daily consumption was 6,583,290 New York gallons.

From experiments made by Mr. Lane in May, 1864, to which I will again refer, we have the means of getting very nearly the proportion of the twenty-four hours' consumption which belongs to the hours of daylight. From these data the flow per hour from Ridgewood reservoir, assumed, for the 17th February, 1863, to be equal to the average of that month, is found to be 346,170 gallons. Allowing as before one fourth of this to be drawn off by the Williamsburgh branch, there remain 259,628 gallons flowing to Brooklyn proper.

To both of these quantities there is to be added the additional flow of this particular day, caused by the working of the pumping engine, taken at 110,000 gallons. From these data the leading quantities in the following table are made up ; the other figures are sufficiently explained in the table.

TABLE ILLUSTRATIVE OF THE PRESSURE AT THE PUMP WELL, DUE TO THE FLOW OF WATER IN THE CITY MAIN AND ITS PUMP BRANCHES ON FEBRUARY 17, 1863.

| POSITIONS OF MAIN. | Length of Main in feet. | Diameter in inches. | Rate of Flow in New York gallons per hour. | Velocity in feet per second. | Calculated Head to produce the given flow, using the formula $h = 0.00046749 \frac{L}{d} (v + 0.397)^2$ |
|---|---|---|---|---|---|
| Ridgewood Reservoir to De Kalb avenue.. | 15,637 | 36 | 456,170 | 2.294 | 17.653 |
| De Kalb avenue to Washington avenue... | 10,425 | 36 | 369,628 | 1.995 | 9.295 |
| Washington avenue to Pump well ........ | 4,585 | 30 | 150,000 | 1.086 | 1.886 |
| Washington avenue to Pump well......... | 205 | 20 | 150,000 | 2.444 | 0.464 |
| | | | | | 29.298 |
| We added 10 feet before for the effect of branches ; add at this stage of the works 6 feet........ | | | | | 6 |
| | | | | | 35.298 |
| This gives 35.298 as the head lost at this date, by calculation. On the 17th February, 1863, the water in the Ridgewood Reservoir stood above tide at........ | | | | | 165.87 |
| | | | | | 130.572 |

We have thus 130.572 as the calculated height to which the water would have risen at the Mount Prospect engine-house during the day hours, with the engine pumping. The actual condition of the water on that day as observed by the Ashcroft gauge, showed between the hours of 8 and 11 A. M., one hundred and twenty-nine and a half feet as the height to which the water would have risen in an open tube, except at 10 A. M., when the height indicated was one hundred and twenty-six feet. For three hours, then, of that day's pumping, it would have stood at one hundred and twenty-nine and a half feet or below it. This is as close a confirmation of any calculation which the nature of the case admitted of, as I could have expected.

I have referred to some observations lately made by Mr. Lane to ascertain the relative consumption of water within the city during the day hours as compared with the night hours, and of each as compared with any average of the twenty-four hours (6 A. M. to 6 A. M.) constituting a day.

The observations were made hourly at the Ridgewood reservoir and the Ridgewood pumping-engines conjointly, between the 24th and 30th days of May, 1864, and from the data thus obtained the hourly flow into the city was calculated. The Mount Prospect pumping-engine was not working during this time.

The details of these observations will, I presume, be given by Mr. Lane in his yearly report to the new Board of Water Commissioners. I give here a synopsis of the results to show the mode by which I obtained in the table last given, the average rate of consumption during the hours of daylight, from the consumption of any given day of twenty-four hours.

TABLE OF ACTUAL CONSUMPTION OF WATER IN THE ENTIRE CITY OF BROOKLYN, BETWEEN THE HOURS INDICATED.

| 1864. | Day, 6 A. M. to 6 P. M., 12 hours. | Night, 6 P. M. to 6 A. M., 12 hours. | Whole Day, 6 A. M. to 6 A. M., 24 hours. |
|---|---|---|---|
| | New York gallons. | New York gallons. | N. Y. gallons. |
| Wednesday, 25 May | 4,654,395 | 2,324,197 | 6,978,592 |
| Thursday, 26 May | 4,558,781 | 3,335,499 | 7,894,280 |
| Friday, 27 May | 4,776,696 | 2,942,048 | 7,718,744 |
| Saturday, 28 May | 5,455,872 | 3,043,120 | 8,498,992 |
| Sunday, 29 May | 3,683,196 | 2,893,866 | 6,577,062 |
| Monday, 30 May | 4,507,287 | 2,428,740 | 6,936,027 |
| Tuesday, 31 May | 5,650,735 | 2,754,459 | 8,405,194 |
| Seven days | 33,286,962 | 19,721,929 | 53,008,891 |
| Deducting Sunday | 3,683,196 | 2,428,740 | 6,111,936 |
| | 29,603,766 | 17,293,189 | 46,896,955 |
| Average for each week day | 4,933,961 | 2,882,198 | 7,816,159 |

Taking the resulting average for one week-day as our guide, it will be found on trial that the hourly average consumption of the entire day of twenty-four hours, if multiplied by 1.262 will give the hourly average due to the *day*-hours between six A. M. and six P. M., and if multiplied by 0.737 will give the hourly average due to the night-hours between six P. M. and six A. M. In other words, if $x$ be the hourly consumption during any twenty-four hours indicated $1.262x$ will be the hourly consumption between six A. M. and six P. M., and $0.737x$ will be the hourly consumption between six P. M. and six A. M.

This is as we find it in Brooklyn in 1864. For other cities it will vary with their conditions, but it is interesting and valuable to have the relations above given even for one city. The population of Brooklyn now exceeds 300,000.

MOUNT PROSPECT PUMPING-ENGINE.

The contract requires that an engine "shall be erected in the engine-house for the Mount Prospect reservoir," "capable of pumping 2,500,000 gallons per day of sixteen hours," and subject to the same standard of "duty" as the Ridgewood engines. The engine erected has on a preliminary trial satisfied the conditions above-mentioned, as will be found fully stated in the report of Mr. W. E. Worthen, who conducted the test trial, printed in the Appendix.

A Cornish engine was evidently the class of engine contemplated in the contract and in that case a well would have been necessary as a measure of safety; a well was accordingly first thought of as a portion of the engine-house works. Without a well the sensitiveness of the Cornish engine would have been dangerously increased.

The engine built is a crank and fly-wheel engine; its pumps are connected directly with the main, and are thus subjected to all the variations of pressure from hour to hour which occur in the city consumption.

During the night the pressure on the induction main within the engine-house is generally thirty to forty feet above what is requisite there; with an open well this surplus pressure would have been lost, but as now arranged the engine receives the benefit of it, to whatever extent it may happen to exist for the time being.

Whether at night or during the day, it reduces by so much the work to be done on the rising main and economizes the fuel necessary to work the boilers. The engine therefore is better adapted now to meet all the varying conditions of the service, than if its position had been changed every three months to correspond with the head prevailing at each interval.

The steam-cylinder is twenty-four inches diameter. Its length of stroke fifty-four inches. The two pumps are each twenty and a quarter inches diameter. Length of stroke forty-one and a half inches.

The boiler is eighteen feet in length; diameter six feet. The dimensions and number of its flues will be found in Mr. Worthen's report, to which the reader is referred for the form and characteristics of the engine and pumps.

The two pumps stand at the same level, but in the case of these pumps the water from the pump nearest the induction main can reach the rising main without passing through the other pump.

There is an air-chamber on the induction main of a capacity of sixty-nine cubic feet, and one of one hundred cubic feet capacity on the rising main. The length of the force main, which is of twenty inches diameter, is 2,052 feet.

### DRAINAGE GROUNDS.

In my report of the gauging of certain streams made in 1856 and 1857, I attempted to give an estimate, at the same time, of the aggregate extent of the drainage-basins of these streams. The only means which I then had for approximating to such an estimate was the published map of Long Island, very unreliable for such a purpose.

A correct survey of this drainage-basin was on many accounts desirable. In 1859 the construction of the upper division of the conduit was in progress under the charge of Mr. Theodore Weston, as assistant-engineer. The stoppage of all masonry work during the winter months leaves the engineering parties in immediate charge of it somewhat at leisure for other duty between that time and the spring.

Taking advantage of this leisure, I directed Mr. Weston, in the fall of 1859, as soon as the laying of masonry was stopped on the conduit works (which usually took place about the end of October) to set his party to make a careful survey of the drainage-basin of the streams from which our supply of water is obtained. This survey was made accordingly, and the results reported to me in March, 1860, but during my absence in Europe, on account of ill-health, some of its details were elaborated, and the publication of the results deferred, in consequence, until January, 1861. The facts ascertained by this survey will be given here.

It is important to bear in mind of this drainage basin, as of all drainage basins, that the amount of water available from it will always be a function, or a fraction of the yearly rain-fall simply, and that beyond the available portion of the yearly rain-fall there is no other source of supply to be looked to. If we draw our supply from a large river, as in the case of Hartford city, Philadelphia, Buffalo, or Cincinnati, the portion taken bears so small a relation to the flow of the particular stream, or the drainage basin in these cases is so largely beyond our wants, as to render any speculation connected with its size superfluous. But in the case of the cities of New York, Boston, or Brooklyn, it is otherwise. In each of these cases, the entire drainage basin of the stream or streams from which the city supply is derived, will require eventually to be utilized, and should be brought within the necessary control in season by special legislation.

The size of each basin with its rain-fall indicates all the water used upon or available from it in any shape during the year. Part of that water is absorbed by the soil and sub-soil, and held in suspension, so to say, to maintain that degree of moisture there which is necessary to the wants of vegetation; part of it is returned to the atmosphere in evaporation, after moistening the leaves of the trees and grasses; part of it flows over the surface directly into the brooks during heavy rain-falls, or, in winter, during the thaws; the residue finds its way under ground into the spaces, fissures, sands, and other retentive earths, whence it slowly escapes, delayed by the friction of these earths, and the economic conditions of the rocks in this respect, reaching the surface again on the valleys in the form of springs, and forming in summer about the sole reliance of the brooks and streams.

From that portion of the rain-fall which escapes into the brooks either directly or indirectly, our present supply is obtained. We use now but a fraction of this portion of the rain-fall; the larger portion runs to waste, but it can be in large part made available hereafter, when desired.

From the survey already adverted to, the drainage-basin of the streams connected with our works was found to measure sixty and a quarter square miles. Within the limits of the same basin there lie the small valleys of Springfield valley and

Foster's brook not connected with our works. The drainage area of these outlying streams, including besides all the unused water-bearing lands to the north of the conduit line, measures 13.41 square miles. Of this outlying portion of the drainage, Springfield brook alone would be worth the cost of ponding and bringing into the conduit.

Our drainage grounds lie partly on the southern slopes of the central ridge of hills already described, but mainly on the plain or prairie extending from Jamaica towards and beyond Hempstead, known as the Hempstead plains.

The hill slopes first mentioned are composed of a clayey alluvial earth, largely intermixed with boulders and but little retentive of water. The plain referred to is on the contrary largely receptive and retentive of water. It consists of a very uniform deposit of clear sand and gravel, with occasionally and very rarely thin veins of clay. The proportion of the drainage grounds which lie on this gravel plain cannot be correctly defined, but it is roughly estimated at forty square miles. The wells which are scattered over this gravel plain enable us to understand the position of its underlying water above tide.

This gravel plain serves two purposes as regards that portion of the rain fall which sinks into it. It retains it as fissures in the rocks retain it in other formations (such as prevail in the Croton Basin), to deliver it slowly to the surface at the brooks or in the bay shores, in the shape of springs, and it filters and purifies it, the gravel and sand acting as a large natural filtering bed.

Of the water found in place over the whole of this gravel plain, but a small portion has been derived from the rain of the current season; the large mass of it has grown to its present height from the gatherings of a series of years.

The permanent shape of this mass deserves attention. It has a pretty uniform inclination towards the southern shore of twelve feet per mile. Upon each low ridge lying between the several streams which cross the Hempstead plain, the inclination of this water roof varies with the width of the ridge and must be steeper than its sea slope, the hindrance to the escape of the water being proportionally less. This permanent shape or regime of the underground fluid grows out of the character of the material in which it is embedded. As much of the rain fall as sinks into the ground flows over the fluid slopes of this underground water shed as upon inclines, the greater part of it thus finding its way into the brooks, and a portion of it escaping between the mouths of the brooks into the neighboring bays. The longer the time occupied by these accessions of rain in reaching the outlets indicated, the greater will be the minimum flow of the brook as compared with its total flow, and the shorter the time occupied, the smaller will be the minimum flow comparatively,—so that in some limestone countries, where the fissures in the rocks are large and free, the minimum is frequently reduced to zero, the brooks becoming dry during some of the summer weeks. In our case the fine sand operates so as to render the flow of the water through it very slow.

Upon the permanency of this regime, so called, the uniformity of our water supply depends, varying only, and that slightly, with the annual variations in the rain-fall.

If we can conceive of an entire year passing without rain, this regime would be changed and lowered, a portion of the water maintaining it oozing out then into the brooks and bays ; and supposing a series of dry years, with no rain-fall, to follow each other, if that were possible, it would gradually sink to the level of the tides.

Under such circumstances the flow of water in the brooks would at length cease, and before it could resume its present position the mass of water withdrawn from the sand plains would have to be restored—the same regime would have to be re-established to secure the prevalence of the same results.

In the same manner if, by artificial drains taking the shape of or performing the functions of new brooks, we were to lower this underlying water and disturb this regime, the amount flowing into the old brooks would become proportionally less ; the head of water, in other words, acting upon all the little springs would be lowered, and their deliveries would be lessened. We might in this way produce a lower regime and appropriate the water forming the difference between the old and the new, but it would only be to arrive at a state of things as a matter of course, in which not more than the same proportion of the rain-fall which settles into the ground would be available to be utilized for the purposes of these works.

I have ventured to say this much upon a question which is debatable in some minds. A more practicable subject in connection with our drainage-basin, will be the consideration of how much of the water of the supply-brooks runs to waste over our dams, and whether a large portion of this cannot be recovered and applied when the wants of the city demand it, before looking beyond the present drainage-basin for increased supplies.

The following table gives the results of careful gaugings, made in September and October of 1856 and 1857, of the brooks then expected to be included in the works. One of these, "Springfield brook," was not taken in by the contractors, their intention being to take in an equivalent of water from the deep cut near the engine-house, by openings left in the masonry of the conduit for that purpose. These openings have not, however, been used, nor is there any water derived thus far from that source.

We know, however, from experiments carefully made for that purpose in 1858, when the depth of water in the pump-well section of the conduit was three and a half feet, that one million four hundred and two thousand gallons of water can be derived from these holes. With five feet of water in the conduit the result would be lessened.

STATEMENT OF THE RESULTS OF GAUGING MADE DURING THE AUTUMNS OF 1856 AND 1857, TO ASCERTAIN THE MINIMUM DELIVERY OF THE PRINCIPAL STREAMS, FROM WHICH THE SUPPLY IS OBTAINED FOR THE BROOKLYN WATER WORKS UNDER THE CONTRACT OF H. S. WELLES & CO.

| NAME OF STREAM. | Date. | Minimum rate of the Stream for twenty-four hours during the period observed. | | | | Minimum flow of Water in the Conduit at the receiving points of the different streams. |
|---|---|---|---|---|---|---|
| | | By Francis' Formula. | By Weisbach's Formula. | By Taylor's Formula. | Mean of the three Formulas. | |
| | | N. Y. gallons. | N. Y. gallons. | N. Y. gallons. | N. Y. gallons. | |
| Hempstead Stream.................. | ............ | 7,651,321 | 8,242,371 | 8,826.148 | 8,239,947 | 8,239,947 |
| Rockville Stream (Pine's Brook)... | ............ | 2,569,320 | 2,758,800 | 2,954,420 | 2,760,847 | 11,000,794 |
| Valley Stream (P. Cornell's)....... | ............ | 2,388,305 | 2,528,275 | 2,707,425 | 2,541,335 | 13,542,129 |
| Clear Stream ...................... | ............ | 739,400 | 780,000 | 834,850 | 784,750 | 14,326,879 |
| Brookfield Stream (Simonson's).... | ............ | 1,878,136 | 1,991,450 | 2,132,700 | 2,000,762 | 16,327,641 |
| Springfield Stream................ | ............ | 634,256 | 673,645 | 721,360 | 676,420 | 17,004,061 |
| Jamaica Stream.................... | ............ | 3,053,792 | 3,271,110 | 3,502,792 | 3,275,898 | 20,279,959 |
| | | 18,914,530 | 20,245,651 | 21,679,695 | 20,279,959 | ...... |

After the completion of the conduit, and before its formal acceptance, the waters flowing through it from the regular delivery of the six brooks of our supply-ponds were repeatedly measured, and found always to exceed the contract quantity of twenty millions, and to exceed decidedly the gaugings of 1856–7. The cleaning out of the ponds would, in part, account for this last condition. The conduit works were accordingly accepted as satisfying, at the time, the contract requirements in this respect.

With the contract quantity of twenty millions of gallons flowing through the conduit, the supply-ponds possess a reserve of seventy millions of gallons in all, which can be drawn upon during any year of extreme low water. Such seasons of unusual drought occur once in twenty or thirty years. With a flow of forty millions daily passing through the conduit, equal to five feet of water there, the reserve referred to would be reduced to thirty-one millions of gallons.

The gaugings given in the above tables were taken during very dry seasons, as represented by the millers. The Flatbush and Jamaica tables of rain-fall somewhat corroborate their opinion, but I place little confidence in such tables unless evidence is at the same time furnished that the observations have been made with great care and with reliable instruments. The tables above mentioned differ very importantly and very irregularly, although the places lie on the same plain and are but nine miles apart.

If we take the summer flow of our supply-brooks at twenty millions of New York gallons, or two millions five hundred and sixty thousand cubic feet per diem—equal to nine hundred and thirty-four millions four hundred thousand cubic feet per annum—and compare it with the area of the drainage-basin, it will be found equivalent to 0.5585 feet, or 6.7 inches of rain-fall. This, of itself, is evidence of a formation unusually receptive and retentive of that portion of the rain-fall which is available for natural storage. It is unnecessary to repeat that the sands of the Hempstead plains have this marked character, being covered by a very light and thin soil, through which the rain finds an easy passage to the sand; it is restored again to the surface by springs, very slowly and very regularly.

Mr. T. F. Bateman mentions the sands of Delaware forest, in Cheshire, England, as yielding, "from measurements made in the summer of 1851, sixteen millions of gallons (Imperial) a day from a tract of country not exceeding thirty-six square miles in extent." This exceeds the minimum flow of our brooks, but it represents, probably, an average of the summer low water, and not its very lowest flow.

The minimum flow of the Croton river may be taken at thirty-seven millions of New York gallons per diem, equal to four millions seven hundred and thirty-six thousand cubic feet.* The drainage-basin of the Croton river measures three hundred and thirty-eight square miles. The minimum flow in this case is equivalent to about two inches of rain. This shows a character of basin upon which, if the rocks receive within their strata as much of the rain-fall as in our case, they must restore it again to the surface much more rapidly and freely than in our case, many of its springs and brooks becoming all but dry in the fall months. More probably the formations in the Croton basin do not store up the same proportion of water, but shed off into the rivers a larger fraction of every heavy rain.

During the lowest stage of the streams, their sole dependence is, of course, upon the springs. The yield of the Croton under these circumstances approximates pretty nearly to that of some English rivers. Mr. Bateman says: "The spring water varies in extreme drought from one quarter of a cubic foot per second for every thousand acres (1.562 square miles) of contributing area—as in the Washbourne, one of the tributaries of the river Wharfe, in Yorkshire—to three quarters of a cubic foot per second from the same area—as in the river Etherow, at the Manchester water works. The spring water of the Rivington Hills, whence the supply of Liverpool is to be obtained, is equal, in the same dry season, to about half the quantity of that yielded by the Manchester districts in proportion to their respective areas. The general lowest yield of these measures in the driest weather, after a long period of drought, is about one fifth or one half of a cubic foot per one thousand acres.

The yield of the Croton basin in such a drought, according to the estimate which I have given of its minimum flow, is about three tenths of a cubic foot per thousand

* Aided by the reserve in the Croton lake, Mr. Schramke makes the minimum equal to thirty-five millions of Imperial gallons, or forty-three millions, seven hundred and fifty thousand New York gallons. Mr. Albert Stein, from gaugings made in September, 1833, reported the minimum as twenty-nine millions, four hundred and seven thousand, six hundred and twenty-two gallons every twenty-four hours. Mr. Horatio Allen, in 1858, made it twenty-seven millions, five hundred and sixty-two thousand, two hundred and eighty gallons (Imperial). The year 1833 is reported as one of uncommon drought.

acres. The minimum yield of the Brooklyn basin exceeds three quarters of a cubic foot (0.77) for the same area.

But in proportion as the yield of the springs during the driest month of the year, is great in any basin, will the water passing off by the streams during the other months be less as compared with a basin of the opposite character. The water which runs to waste over the Croton dam is very much greater proportionally than the water which runs to waste over our dams. They have, therefore, a larger proportion of unapplied water to store up at need than we have, and the Reservoir works required to economize their surplus, will be proportionally more expensive than in our case.

We come back to the consideration of the amount of water wasted over the Brooklyn supply dams, over and above the twenty millions of natural flow at the lowest stage. This is a variable depending on the rain-fall, the lowest value of which will govern our present calculations.

Of the rain-fall of any year it has already been stated that a certain portion is expended in evaporation, vegetation, and absorption, the absorption referred to being that dampness or moistened state of the soil or earth near the surface, which is necessary to the health of the animal and vegetable life residing in that soil.

The remainder in part passes through this upper surface to the springs of whatever character, and in part is shed directly into the brooks from the surface of the ground.

The portion of the rain-fall applicable to vegetation, evaporation, and soil absorption, is not now considered to vary sensibly with the annual rain-fall, but is supposed by those who have been able to give the subject due consideration to be very nearly a fixed quantity; being from fourteen to sixteen inches of the yearly rain-fall.

Of these tables of rain-fall, viz., the Flatbush table, the Fort Hamilton table, and the Fort Columbus table (see Appendix), the lowest year of the Fort Hamilton series, 1849, corresponds very nearly with the same year in the other two series.

| | Inches. |
|---|---|
| The Fort Hamilton record gives for 1849 | 29.75 |
| The Flatbush " " " | 32.47 |
| The Fort Columbus " " " | 31.14 |
| | 93.36 |
| Mean for 1849 | 31.12 |

The lowest year of the Flatbush series (thirty-two years) is 1845, 32.14 inches. The lowest year of the Fort Hamilton series (fourteen years) is 1849, as above mentioned. The lowest year of the Fort Columbus series (nineteen years) is 1836, 27.57 inches, not given in the Fort Hamilton series, but given in the Flatbush series at 43.89 inches, and, therefore, probably erroneous. The mean given above for 1849 is the lowest mean of any year in these three tables, and we may, therefore, take it in the absence of better data as our low-water standard for the rain-fall.

| | |
|---|---|
| I will take, however, 30 inches as a safe standard for a very dry season in our drainage basin.......................... | 30 |
| Allow for evaporation, vegetation, and such absorption as does not reappear in springs................................ | 15 |
| | 15 |
| Leaving 15 inches as applicable to springs and floods, as the water in fact which flows off by the six brooks whose waters belong to our works. | |
| Of this portion the minimum flow of the brooks (which is all that our works as now arranged can obtain), during the driest month of the year, equals twenty millions New York gallons, and this is equivalent to 6.7 inches of rain-fall.... | 6.7 |
| | 8.3 |

There remains then an equivalent of 8.3 inches of rain-fall wasted, over and above the twenty millions gallons which can be taken without storage. This is equal to twenty-four millions seven hundred and fifty thousand New York gallons per diem. During ordinary years, when the rain-fall generally exceeds forty inches, this surplus quantity would much exceed its computed allowance for a dry season. It is evident, then, that from the limited drainage basin of the present works forty millions of gallons can be gathered and depended on without looking farther, if we can husband the water which runs to waste.

This can probably be done in one of two ways which naturally suggest themselves.

But we must first consider the amount of storage to be provided for. Every river basin seems to have its individual character in this respect, and to be able intelligently to understand hereafter the necessities of our situation, one of the largest of our streams, the Hempstead brook for instance, should be gauged daily, at least every four hours for two years, estimating at the same time its daily delivery into the conduit, and observing carefully the rain-fall during the same period. This would sufficiently indicate the character of the other streams, and would enable the Engineer who may be called upon to study the matter, to understand pretty nearly the extent of reservoir storage necessary to economize the surplus waters of the basin.

In the absence of such data as the observations above indicated would give, I will assume that the flow of the supply-brooks during the months of January, February, March, April, and May, amounts to from two to four times the flow at their lowest stage (twenty millions gallons), say two and three fourths times that flow; that in June, July, and August, it amounts to one and a half to two times that flow, say an average of one and three fourths times; and that during the months of September, October, November, and December, it varies from one to one and a half times that flow, say an average of one and one fourth times.

9

The surplus waters to be preserved have been shown to be equal to a per diem of twenty-four and three fourths millions of gallons during a season of very low rain-fall.

To secure the daily supply of forty millions gallons, which the basin is amply competent to give; in other words, to lay by from the winter and spring waters enough to make up the deficiencies of the summer flow upon this standard, there would eventually be required reservoir storage equivalent to two thousand seven hundred millions New York gallons—equal to a contributing capacity from the said reservoir of six millions gallons average per diem for ninety days, and of eighteen millions gallons average per diem for one hundred and twenty days—quantities which correct data would probably reduce.

Two modes have been alluded to of meeting this storage. The first is the usual one of building reservoirs on the streams. This need not be done on every stream. If such reservoirs were built (as needed) on Jamaica brook, Valley brook, Rockville brook, and Hempstead brook, the other two brooks could be so controlled as at all times to deliver all their waters into the conduit, the deficiency being made up from the other streams by an intelligent supervision of them.

The valleys alluded to are, however, shallow and not well adapted to the storage of so much water without covering very large areas, involving the acquisition of much land, and the removal of proportionally large quantities of vegetable deposit or muck. As a question of cost, therefore, it may turn out more economical to build one large reservoir situated on the neck of land lying between Hempstead pond and Pine's pond or elsewhere. To meet the reserve above mentioned, this reservoir with twenty-four feet depth of water would have a water surface equal to three thousand eight hundred feet by three thousand eight hundred feet. Its bottom would have to be situated say four feet above Hempstead pond, necessitating, therefore, engine power to pump into it from the waters of Pine's pond and Hempstead pond the necessary storage water. The whole of the water of the other streams would in that case be graduated into the conduit, making up from this storage reservoir the deficiency. The reservoir would be built in two parts.

The pumping in this case will appear objectionable, but the height to be pumped is small and the cost of pumping will be small as compared with Ridgewood. A considerable portion of the water which finds its way into Smith's pond may be drawn into the pump-well, and if we consider the compactness of the scheme, the small amount of land required as compared with the other, and the depth of water in this case as compared with the shallower waters of any reservoirs built on the streams, it may turn out as economical and more manageable than the mode first mentioned.

I mention it only as another way of securing the necessary storage. The best mode can only be ascertained after the usual correct surveys and calculations have been made and the whole matter well studied, a necessity which will not probably occur for many years yet.

Having thus described the works as briefly as I have been able to do consistent with the desire to have them understood, I beg to say in this place how much I feel myself beholden to all the gentlemen of the Board of Water Commissioners, for the liberal manner in which they have uniformly sustained me in the prosecution of duties which were in many respects unusually arduous. It would have been easy for them to have embarrassed me in the selection of assistants and inspectors, upon whose zeal and faithfulness the right construction of the various works depends ; but they have always encouraged me to secure the most reliable men, and always been eager without favor to get rid of any one, and such were very rare, who proved himself unfaithful or incompetent. The perfection of any work of this kind obviously very much depends on the presence of such a spirit in those having the control of it.

I have also to acknowledge my obligations to A. W. Craven, Esq., who acted as Consulting Engineer, for access at all times to the plans of the Croton works, and for the frequent aid of his judgment.

I have also to thank the Assistant-Engineers for their attention and faithfulness to their several trusts, of which the works afford sufficient evidence.

The Assistant-Engineers were Mr. Moses Lane, principal assistant.

Mr. Samuel McElroy, in charge of the construction of the Ridgewood reservoir, and of the engine-house, and pumping-engine No. 1.

Mr. Theodore Weston, in charge of the construction of the conduit.

Mr. Joseph Bennett, in charge of the surveys, and the construction of some of the supply ponds.

Mr. Joseph P. Davis, in charge of the construction of the foundations of Ridgewood engine No. 2, and aiding Mr. Weston on the Mount Prospect reservoir.

Mr. Joseph O. B. Webster, in charge of the drawings of the city pipe distribution, and of the pipe returns and estimates.

Mr. Webster, Sr., in general charge of the out-door work of the pipe service at the foundries and in the city.

Mr. Frederick Budden, draftsman and architect, in charge of all working details of the engine-houses and buildings on the works, all which were built according to his designs, with the exception of the Ridgewood engine-house.

The inspectors of the pipe castings, resident at the foundries, rendered also most important service in securing for us a superior class of castings. These were :

Mr. F. B. Miles.

Mr. Jno. H. Rhodes.

Mr. S. R. Probasco.

Mr. John Avery, Jr.

Mr. J. D. Mercer.

The total expenditures to date by the Board of Water Commissioners for construction have been as follows:

| | | |
|---|---|---|
| The original contract price | $4,200,000 00 | |
| Additional cost for construction of a conduit instead of a canal and other outside expenses | 498,495 58 | |
| Total expenditure through H. S. Welles & Co | | $4,698,495 58 |
| Additional coal building and railroad track at Ridgewood engine-house | 5,000 96 | |
| Instruments and office furniture | 2,619 76 | |
| Discount on bonds | 22,849 65 | |
| Engineering, superintending, office expenses, and contingencies | 78,711 47 | |
| | | 109,181 84 |
| The extension of the distribution water pipes and laying, engineering, superintending and contingencies | 471,543 41 | |
| Additional lands, &c. | 5,405 38 | |
| | | 476,948 79 |
| Total expenditure by the Board of Water Commissioners for construction | | $5,284,626 21 |

It becomes me here to say of H. S. Welles & Co., the Contractors, now that our relations are terminated, that, although we frequently differed in matters of detail, their conduct in the fulfilment of their contract was, on the whole, liberal and creditable to them as constructors. Their general sound knowledge of what constituted good work, and their support of the Engineers in requiring it rigidly of their sub-contractors, materially simplified the duties of superintendence.

# SEWERAGE WORKS.

# SEWERAGE WORKS.

By an act of the legislature, passed 15th April, 1857, the Board of Commissioners for the construction of the water works, already described, was authorized to prepare a system of sewerage for the city of Brooklyn, and to proceed with the construction of sewers under that system.

On the 13th July, 1857, I was appointed by the Board the Engineer to prepare and organize such a system in conformity with the provisions of the act; but as my duties in connection with the construction of the water works fully occupied my time, the Board authorized me to procure the services of Julius W. Adams, Civil Engineer, to study out and prepare the necessary plans. Soon after Mr. Adams entered upon his duties, I resigned as Engineer, and he was appointed to that position. The plans of the sewerage districts as now established are the result of his labors, and what I have to say in explanation of them will be mainly in accordance with the views expressed in his reports and illustrative of the principles which governed him.

In anticipation of such duties devolving on me, I had previously collected, from English sources principally, such information with regard to the drainage of cities as had been published in parliamentary reports, and elsewhere, and I had corresponded with some English engineers of large experience in such works, particularly with Mr. Robert Rawlinson, who very liberally gave me the benefit of his views and furnished me with valuable information as to the character of the sewerage works then recently constructed for many of the English cities. To him I feel under great obligations for his exceeding courtesy in this respect. This correspondence, together with the reports alluded to, were handed over to Mr. Adams on his entering upon his duties.

The establishment of works to insure a liberal and regular supply of water for any city, involves the establishment as well of the sewerage works necessary to carry off that water after it has been used. Without the last, a free use of the water would for obvious reasons be impracticable.

When the supply of water is deficient or limited, that is to say, when (as was usual thirty years ago) it has to be procured from street wells or purchased from water-peddlers, the amount used is but a small fraction of what is considered necessary for domestic purposes now; so small that cesspools placed in the back yards of the dwellings were competent then to receive it after use and to pass it into the soil by percolation; but such percolations, and the percolation from the vaults of the privies situated in the same back yards, spreading in each case, year by year, over larger circles of the subsoil, have invariably sooner or later reached the water of the wells from which the inhabitants receive their daily drink, and doubtless have been long offensive to health before they have become distinctly perceptible or offensive to the taste. In many cases the taste of individuals, as in the case of sulphur and other offensive springs, has accommodated itself to the changed character of the water, and has afterwards found a purer water somewhat insipid in comparison. The soil under this system becomes after a while saturated for a considerable depth with the impurities incident to human life; the rain-waters, upon which all wells depend for their current supplies, necessarily reach these wells after passing through more or less of this earth saturated with such impurities, and they imbibe in their course a portion of its poisonous qualities. These results, slowly believed in and generally suffered to generate malignant forms of disease before being admitted, have led equally to the demand for a liberal supply of the purer water of the country and for such a change in the domestic arrangements as regards water-closets and all refuse water of the family, as shall remove it promptly not merely from off the premises, but out of the city and beyond any influence for evil upon its atmosphere. The necessities of overcharged vaults have led in some old cities to the periodical removal of human ordure, but the process poisoned the atmosphere more sensibly at those times, and left the lungs to inhale what formerly had been received into the system mainly through the stomach. These remarks may appear irrelevant in our day, when the dependence of health on cleanliness of habits as well as of person is so generally admitted, but the admission will avail but partially unless the effort to attain the desired end is made continuous and not intermittent:—the best of our sewerage works fail in several important particulars of that perfection which is necessary to render them entirely innocuous; they will, however, reach it step by step if the public mind remains watchfully sensitive to the truth that a pure atmosphere in-doors and out of doors is one of the essential elements of health which must always be of difficult attainment in a crowded city.

The total length of the sewers existing in Brooklyn previous to 1857 was $5\frac{55}{100}$ miles. The larger ones were situated as follows:

Raymond street, Park to De Kalb avenue.
Navy street, Prospect to Tillary street.
Fulton street, East river to Clinton street.
Warren street, East river to Court.
Hamilton avenue, East river to Rapelyea street.
Tillary street, Raymond to Bridge street.
Union street, Hamilton to Columbia street.
Smith street, Atlantic to Warren street.
Across the Navy-yard and City Park to Raymond street.
Navy street, De Kalb to Flatbush avenue.

These sewers were evidently built to relieve certain depressed portions of the city of the accumulations of water which must have found place there during heavy rains. They were not built to relieve the house drainage, which was at that time otherwise disposed of, with the exception of such portions of it as were thrown out in front into the gutters. This condition of things necessarily continued until the introduction of water into the city made a change for the better practicable as well necessary.

The sewers referred to were made large enough to admit of their being entered by men, and their accumulations cleaned out when they became offensive or otherwise objectionable.

It will be convenient to look back to the practice of English cities previous to 1850, for we have been largely enlightened and benefited by their experience and by their disputes on the subject of city drainage.

Previous to 1850, or thereabouts, the sewerage drains of many English cities had been established so as to receive the house sewerage as well as the street sewerage, or rain-fall, but the water supply of the same cities being then deficient in the quantity used per head of population, as compared with our present habits, the resulting fluid sewerage of dwelling-houses was not sufficient in quantity to carry off the heavier matter which accompanied it. This last, therefore, when it reached the sewer deposited itself there in part, and accumulated upon the bottom until carried away by the waters of heavy rains, or removed by manual labor. These accumulations frequently remained long enough to become putrescent, and by their decomposition, exceedingly offensive to health and sense. The exaggerations of sickness which grew out of this state of affairs are matters of history, not likely to have place again to the same extent in our times. The sewers under this system were all built of sufficient size to admit of their being cleaned out at intervals by manual labor.

The introduction into the same cities, and gradually into all important cities of more abundant supplies of water, admitting of its free use for the luxuries as well as the mere necessities of domestic life, paved the way for a change in the modes of sewerage previously prevalent.

This change consisted in the introduction of smaller sewers, the sizes of which were governed solely by the amounts of sewerage and rain water to be passed at any given point; their enlargement beyond the requirements above mentioned was considered unnecessary, and in other respects objectionable.

The opinions of English engineers, as regards this innovation, were very various and very contradictory. The general correctness of the views of those who originated it is now sufficiently established by experience.

In the change of system the application of pipe-sewers was incidental;—the controversy sometimes took the shape of pipes against bricks, enlisting in opposition all the interests connected with the latter, but the true controversy lay between small sewers for drainage only, and large sewers for drainage and entrance also.

The smaller sewers taking the shape within certain limits of pipe-sewers originated in the motives of economy, but their continued application was the result of other advantageous conditions which will appear as we proceed.

The ruling principle insisted on, and we presume generally admitted now, is this; that each day's sewerage of each street, and of each dwelling, should be removed from the city, if at all practicable, on the day of its production; that it should pass off before decomposition begins; that it should not be allowed to settle and fester in the sewers, producing those noxious gases which are so prejudicial to health. This principle requires that the sewerage matter of any one day shall not be found within the limits of the city on the next day.

To attain this desirable end, the refuse fluids of dwellings and factories must be sufficient in quantity to float and carry off the heavier matters of sewerage. A liberal supply of water must be within reach of the inhabitants, and their habits must incline them to use it liberally. The water supply of cities in our time is, at least double what was considered necessary thirty years ago, and this favors largely the change in the modes of sewerage now prevalent, and mainly sustains it. If the rate of inclination of a sewer is not flatter than one foot in four hundred and forty, the experience of Brooklyn and of other cities equally well supplied with water, shows that the fluid domestic sewerage of any street is competent to carry off daily all the heavier matters of sewerage, and to keep the drains free and clean, provided that the form of the sewer is such as to concentrate these sewerage waters as much as possible.

If the sewer is made unnecessarily large, the depth and velocity of the fluid passing through it are proportionally reduced, and its ability to keep itself clear will be confined to high rates of inclination. It is important, therefore, to have it as small as the service to be required of it will admit.

As regards the necessity of entering sewers, which was dwelt upon so much ten years ago, the successful use of the small sewers shows that this necessity does not exist; they have been found safe and effective when properly built, and less troublesome as regards cleansing than the old sewers.

Their most important advantage, indeed, grows out of what was contended to be their great defect, viz.: their incompetency to receive or admit of sewerage deposits or accumulations. They presuppose the daily removal of all sewerage matters; and this condition is always necessary to their efficient operation.

For inclinations less than one in four hundred and forty some flushing will be occasionally indispensable; but to keep the small sewer clean, less flushing will be requisite, than for a sewer needlessly large and therefore admitting of and inviting deposit. In Brooklyn the rates of inclination available for the small sewers have rendered resort to flushing unnecessary. Towards the outlets of the sewers, upon the river, the inclinations are necessarily less in many instances than the limit above mentioned, and in such cases flushing by means of tide-gates may hereafter be found essential, though thus far the difficulty has not been experienced.

In Chicago, where pipe-sewers of twelve inches diameter are used with an inclination of one in five hundred, a tank placed on wheels has been used wherewith to flush them occasionally. Where, however, the leaders of the dwelling-houses are connected with the house-drains this precaution has not been found necessary. I have this information from Mr. E. S. Chesbrough, the City Engineer of Chicago, whose researches and experience in sewerage matters are well known, and have been always very obligingly communicated.

In cities where a sufficient rate of inclination cannot be obtained, resort must be had to flushing to insure the free action of the drains. The city of Brooklyn is so happily situated in this respect as to render these artificial aids unnecessary.

The house-sewerage spoken of above is to be understood as not including any rain-fall. In this city, and in all cities within the limits of grade mentioned, the daily flow through its drains must be able to keep them free, independent of any aid from rain.

The point being conceded that the greater the concentration of the minimum flow (in other words, of the sewerage proper) the greater will be its carrying and cleansing power in a sewer of any given inclination, and the concentration being greater as the sewer is smaller, the minimum sizes of sewers practicable come next under consideration.

The conditions governing the size of a sewer at any one point are these:

1. The amount of the sewerage proper passing that point.

2. The amount of rain-fall governing that point.

3. The inclination of the sewer.

When these sewerage works were planned and commenced we were dependent entirely upon English experiments and English experience for any data in regard to the amounts of sewerage per dwelling-house, or per city acre, to be provided for, as well as for the amount of any given rain-storm which reaches a sewer within a given time. We are not in a better condition now as regards any definite experiments made since by ourselves, but we have the satisfaction of knowing that Mr. Adams' application of these data has produced successful results, that our sewers, small as they were thought to be when adopted, have been proved to be abundantly large, with but one small exception, to which I shall advert again.

Experiments made upon certain of the London sewers showed the sewerage proper there to be 4.8 cubic feet per head per diem in the poor districts, and eight cubic feet per head per diem in the wealthier districts. This was understood to include the discharge from manufactories and everything, except the rain-fall. Take eight cubic feet per head per diem as a maximum.

The extremes of rain-fall mainly determine the sizes of sewers, the household sewerage becomes a very small item in comparison.

A rain-storm of one inch rain-fall in one hour has been taken by Mr. Adams as the maximum of rain to be provided for. This occurs very rarely, and experiments made elsewhere show that of the rain falling during a heavy storm in one hour not more than one half of its amount, rarely as much as one third, finds its way to the sewers within the hour. Its progress from the roofs of houses to the leaders and drains, and from the surfaces of the streets to the gutters at their corners, consumes necessarily a certain amount of time, so that the water will always be found flowing in the streets for a certain time after the rain has ceased to flow. A considerable amount of the water passes the gully openings and flows on in the gutters to the lower parts of the city, some of it finding its way into the river independent of the sewers, and all of it consuming time in the gutter channels.

Mr. Adams estimated that from these influences, where one inch of rain fell within one hour, not more than one half of the resulting quantities of water reached the sewer within the same hour; one half inch of rain in one hour became therefore the maximum of rain-fall as regards sewer dimensions.

In searching for a simple formula which shall embody with sufficient accuracy the conditions of the drainage question, and approximate at the same time to the returns of English experimentalists, I have taken the rain-fall entering the sewers from the street as equivalent to one half inch per hour—from the roofs of the houses as equal to two thirds of an inch per hour—the back yards as equal to one quarter inch per hour:—the result corresponds very nearly with the principles kept in view by Mr. Adams.

The formula of Mr. Bazalgette, given by him in evidence in 1852, uses the area of each case in acres as the basis of the sewerage dimensions. I propose to make the length of street the basis, as of simpler application, though its relation to the acres of area will be given also.

It will be convenient to follow out this calculation now.

The streets in the city of Brooklyn are generally sixty feet in width; the lots on either side of them are generally one hundred feet in depth. Where the streets are of less width the formula will err in giving a little larger dimensions than would be strictly necessary; where they are of greater width the calculated length of street would be increased accordingly and the strict percentage of the result applied to it.

One hundred feet in length of a street will be taken as the unit.

We have then for each one hundred feet in length of street, a width of $100+60+100=260$ feet, equivalent to an area of twenty-six thousand square feet.

Let the houses average 24 feet in width each and allow 10 inhabitants to each house. Eight cubic feet per diem having been taken as the maximum house sewerage per head per diem, we have 80 cubic feet per diem for one house, and for the houses (4.16) on both sides of the street on 100 feet of its length, $666\frac{2}{3}$ cubic feet in 24 hours—equal to an average of 27.77 cubic feet per hour or 0.00773 cubic feet per second; but the flow during eight hours of the day has been found to be equal to 1.5 times this average flow.

| | CUBIC FEET. |
|---|---|
| The maximum therefore becomes per second...................... | 0.0116 |
| The area of 100 feet of street being 6,000 square feet, one half inch of rain-fall in one hour is equal to per second.............. | 0.0694 |
| Take the houses as 60 feet in depth on each side, giving an area of roof of 12,000 square feet, which. at two thirds inch of rain-fall per hour, equals per second............................... | 0.1620 |
| There remains 40 feet in depth of back yard on each side, equal to an area of 8,000 square feet, which at one quarter inch rain-fall per hour, is equal to per second................... | 0.0463 |
| | 0.2893 |

We have, then, to provide for a flow of 0.289 cubic feet per second from every hundred feet of street, say in round numbers three tenths (0.3) of a cubic foot per second.

Founded on this standard the following table exemplifies the maximum rates of sewer flowage for given lengths of streets. The corresponding space in acres is added.

| LENGTH OF STREET. FEET. | MAXIMUM DISCHARGE IN CUBIC FEET, PER SECOND. | CORRESPONDING AREA, IN ACRES. |
|---|---|---|
| 100 | 0.3 | 0.5968 |
| 500 | 1.5 | 2.98 |
| 1,000 | 3 | 5.96 |
| 1,500 | 4.5 | 8.95 |
| 2,000 | 6 | 11.93 |
| 3,000 | 9 | 17.90 |
| 4,000 | 12 | 23.87 |
| 5,000 | 15 | 29.84 |
| 6,000 | 18 | 35.81 |
| 7,000 | 21 | 41.78 |
| 8,000 | 2 4 | 47.75 |
| 9,000 | 27 | 53.71 |
| 10,000 | 30 | 59.68 |
| 11,000 | 33 | 65.64 |
| 12,000 | 36 | 71.61 |
| 13,000 | 39 | 77.58 |
| 14,000 | 42 | 83.55 |
| 15,000 | 45 | 89.52 |

The simplest form of Mr. Prony's formula for the flow of water in pipes, as given by Mr. Hughes, is $V = 26.79 \sqrt{D S}$, for measures in metres; when the measures are in English feet the expression becomes $V = 48.49 \sqrt{D S}$. Say, $48.5 \sqrt{D S}$.

This simplicity, which is of great convenience in ordinary practice, is not, of course, obtained except at some sacrifice of accuracy; but provided that the error lies on the right side, extreme accuracy is not so much desirable in this case as convenient application, and it is not obtainable at present in this relation.

As a basis then to obtain a formula which shall give the diameter of drain due to the maximum discharge for any given point, I will take as above,

$$V = 48.5 \sqrt{D S}$$

in which

$V$ is the velocity in feet, per second.

$D$ the diameter of the drain or sewer, in feet.

$S$ the sine of the inclination, or $\frac{h}{l}$ in feet, $h$ being the fall in feet upon any given length $l$.

Call $m$ the discharge in cubic feet per second, and the area being $.7854\ D^2$, we have $\frac{m}{.7854\ D^2} = V$

and

$$\frac{m}{.7854\ D^2} = 48.5 \sqrt{D S}$$

whence $m = 38.08\ D^2 \sqrt{D S}$

and $D = \left(\frac{m^2}{1450}\right)^{\frac{1}{5}}$

In the flow of drinking water through pipes, to which Mr. Prony's researches and formulas apply, the velocity of the water in the pipe is greatest at the source, and decreases with the consumption, or, practically, with the distance from that source.

In the case of sewerage pipes, the case is somewhat the reverse—the amount of sewerage water in the pipes is least at the source, and increases with the contributions of every house-drain and of every contributing branch from the cross-streets of the particular basin. Under the same rate of inclination, the greater the mass of water—provided that it is equally concentrated—the greater will be the velocity. The increase of velocity, however, due to this cause, though sensible, would be but trifling were the flowage of the contributing branches to enter the sewer at about the same rates of velocity as that prevailing in the main sewer at the points of entrance; but during rain-falls they enter the main sewer at much higher rates of velocity, producing a movement there exceeding the velocity, due simply to the rate of inclination of the sewer. Hence the discrepancy between the calculations by well-established formulæ for pure water service, and the actual velocities of sewerage water during rains, as ascertained by experiment, and first dwelt upon in the English reports.

During the rain-storms the water from the roofs of modern houses passes, by means of a tin leader, into the house-drains, with a total fall of from thirty to forty feet, broken, in part, by the bends of the pipe, but nevertheless reaching the street sewer with a high velocity. In older houses, with sharp roofs, one half of the roofage water is delivered into the street, the back half reaching the house-drain in the manner mentioned.

In the case of the street, the storm-water passes into the gullies at the corners of the streets, and thence, with a fall of not less than ten feet in thirty feet of length, is delivered into the street sewer. The velocity of the water entering the sewer must in this case also very distinctly exceed the rate of flow prevalent in the sewer and must increase it.

These are the influences which produce a greater rapidity of movement within the street sewers than what would be due to the rate of inclination simply.

It will be remembered that, although the sewerage proper has to carry along with it a certain amount of solid matter, which would necessarily render its motion less rapid than that of pure water, yet that during rain-storms this material being but a very small fraction of the whole flow, becomes unimportant.

To meet the velocities applicable to sewerage water, the formula last obtained must be modified, and this can be done by changing the root of the general expression for the required diameter.

Mr. Roe's tables furnish me with a means of making this modification. They give the results of his experiments and of his judgment, and in the United States, so far as I know, have been taken as indicating the least dimensions of sewers which should be used for any given area of city drainage. In Brooklyn they have been used more as a check than as a guide, the diameters of the Brooklyn sewers having been always made by Mr. Adams intentionally in excess of Mr. Roe's sizes. The opposition to pipe-sewers, and to what were then called small sewers, and the belief by many of their insufficiency, naturally led to the adoption of sizes greater than what would have been considered necessary under a different state of the public mind.

The following is Mr. Roe's table as given in the report of the General Board of Health, London, 1852 :

TABLE *showing the quantity of covered surface from which circular sewers (with junctions properly connected) will convey away the water coming from a fall of rain of one inch in the hour, with house-drainage, as ascertained in the Holburn and Finsbury Divisions.*

DIAMETERS OF PIPES AND SEWERS IN INCHES.

| | 24 | 30 | 36 | 48 | 60 | 72 | 84 | 96 | 108 | 120 | 132 | 144 |
|---|---|---|---|---|---|---|---|---|---|---|---|---|
| | Acres. | Acres. | Acres. | Acres. | Acres. | Acres. | Acres. | Acres. | Acres. | Acres. | Acres. | Acres. |
| Level........................ | 38¾ | 67¼ | 120 | 277 | 570 | 1,020 | 1,725 | 2,850 | 4,125 | 5,285 | 7,800 | 10,100 |
| ¼ inch in 10 feet, or 1 in 480.... | 43 | 75 | 135 | 308 | 630 | 1,117 | 1,925 | 3,025 | 4,425 | 6,250 | 8,300 | 10,750 |
| ½ inch in 10 feet, or 1 in 240.... | 50 | 87 | 155 | 355 | 735 | 1,318 | 2,225 | 3,500 | 5,100 | 7,175 | 9,550 | 12,400 |
| ¾ inch in 10 feet, or 1 in 160.... | 63 | 113 | 203 | 460 | 950 | 1,692 | 2,875 | 4,500 | 6,575 | 9,250 | 12,300 | 15,950 |
| 1 inch in 10 feet, or 1 in 120..... | 78 | 143 | 257 | 590 | 1,200 | 2,180 | 3,700 | 5,825 | 7,850 | 11,050 | 14,700 | 19,085 |
| 1½ inches in 10 feet, or 1 in 80.. | 90 | 165 | 295 | 670 | 1,385 | 2,486 | 4,225 | 6,625 | .... | ...... | ...... | ...... |
| 2 inches in 10 feet, or 1 in 60.... | 115 | 182 | 318 | 730 | 1,500 | 2,675 | 4,550 | 7,125 | .... | ...... | ...... | ...... |

I quote from Mr. Roe's account of the formation of this table:

"The table is formed from results obtained from observations extending over a period of twenty years in the Holborn and Finsbury divisions.

"In some instances the observations were carried on during the whole period of heavy rains, being commenced as each storm began and continued until the effect had ceased in the sewers, the depth of water being taken every five minutes, and the velocity of the current repeatedly noted at every depth.

"In some instances the observations were continued day and night for several months in different years, and in others they were conducted day and night for two years; rain-gauges being kept to ascertain the depth of rain that fell.

"Every junction, whether of sewer or drain, should enter by a curve of sufficient radius. All turns in the sewer should form true curves, and as even in these there will be more friction than in the straight line, a small addition should at curved points be made to the inclination of the sewer."

The amount of water reaching the sewers from a rain-fall in a given time, will be the greatest on the steepest streets; "for instance, in one case that came under my notice, when the general inclination of the surface of the streets was about 1 in 20, the greatest flow of water from a thunder-storm came to the sewer at the rate of one third more than it did to a sewer draining a similar fall of rain from an area with a general surface inclination of 1 in 132."

"It should be stated that a margin has been left even for an inch of rain."

Although Mr. Roe takes a rain-fall of one inch in the hour as his standard in the above table, his own experiments satisfied him that this amount of water could not reach the sewers within that hour, and the sizes he gives are expressed as sufficient *to convey away the water* coming from that rate of rain-fall.

The formula already given was

$$D = \left(\frac{m^2}{1450\ S}\right)^{\frac{1}{6}}$$

When this is made applicable to acres it becomes

$$D = \left(\frac{n^2}{5804\ S}\right)^{\frac{1}{6}}$$

$n$ standing for acres.

When the last form of formula is altered in its root to $D = \left(\frac{n^2}{5804\ S}\right)^{\frac{1}{6.4}}$ it will be found to approximate very nearly to the diameters of sewers given in Mr. Roe's tables, for all the areas there given, except those where the pipe or sewer has no inclination and is termed level. In effect no such case exists; a certain portion may be built level, as it sometimes will be for a short distance near its outlet, but the surface of the water flowing through that sewer is not level, otherwise it would not flow.

To simplify the root of the formula, and to insure its always giving results somewhat in excess of Mr. Roe's sizes, I give it the following form:

$$D = \left(\frac{n^2}{5804\ S}\right)^{\frac{1}{6}}$$

where it is desired to apply it to the areas in acres—and where it is desired to apply to the cubic feet of discharge at any given point governed in that case by the length of streets delivering to that point as heretofore explained, the formula becomes

$$D = \left(\frac{m^2}{1450\ S}\right)^{\frac{1}{6}}$$

using $\frac{H}{L}$ instead of $S$, $\frac{H}{L}$ being the fall in feet for any given length $L$, these formulæ become $D = \left(\frac{n^2 L}{5804\ H}\right)^{\frac{1}{6}}$, (No. 2); and $D = \left(\frac{m^2 L}{1450\ H}\right)^{\frac{1}{6}}$, (No. 1).

Mr. Joseph W. Bazalgette, Engineer of the London Drainage Works, in his evidence, January, 1857, gives the following as the formula employed by him, "in determining the size of drainage out-fall sewers," "$\frac{3 \log. A + \log. N + 6.8}{10}$ = log. of diameter of sewer in inches required to carry off one inch rain per hour."

$A$ being acres to be drained.

$N$ length in feet in which sewer falls one foot.

This differently expressed is equivalent to

$$D = (A^3 \times N \times 6309574.)^{\frac{1}{10}}, \text{ (No. 3).}$$

$D$ being the diameter in inches.

Preferring to use a formula which I can trace to its origin, I have taken the trouble to work out the other. In the application of either, the engineer will always give due consideration to any peculiarities of climate or details calculated to produce data sensibly different from those assumed here as maxima. In some climates where the rain storms are more violent than with us, it may be necessary to increase the allowance for rain fall. The formula suggested, however, by an appropriate alteration of its root simply, can be made to meet a rain-fall considerably in excess of the amount indicated in the data used by me to obtain it, and also in excess of Mr. Roe's allowances.

In the following table I have applied it to a portion of the areas given in Mr. Roe's tables, to show its results as compared with the sizes given by him. I give also the sizes for the same areas as calculated by Mr. Adams, from Mr. Bazalgette's table.

| MR. ROE'S DIAMETER IN INCHES FOR THE AREAS IN ACRES UNDERMENTIONED. | | 24. | 30. | 36. | 48. | 60. | 72. | 84. | 96. | |
|---|---|---|---|---|---|---|---|---|---|---|
| | 1 in 480 | 43 | 75 | 135 | 308 | 630 | 1117 | 1925 | 3025 | Acres. |
| Calculated size, by formula No. 2.... | ....... | 27.7 | 33.4 | 40.6 | 53.5 | 67.9 | 82.2 | 98.5 | 114.5 | Inches. |
| " " " " No. 3.... | ....... | 27.4 | 32.4 | 38.66 | 49.5 | 61.37 | 72.9 | 85.8 | 98.2 | " |
| | 1 in 240 | 50 | 87 | 155 | 355 | 735 | 1318 | 2225 | 3500 | Acres. |
| Calculated size by formula No. 2.... | ........ | 26 | 31.3 | 37.9 | 50 | 63.7 | 77.4 | 92.1 | 107.1 | Inches. |
| " " " " No. 3.... | ........ | 26.8 | 31.6 | 37.6 | 48.2 | 60 | 71.4 | 83.6 | 95.8 | " |
| | 1 in 160 | 63 | 113 | 203 | 460 | 950 | 1692 | 2875 | 4500 | Acres. |
| Calculated size, by formula No. 2.... | ........ | 26.2 | 31.9 | 38.8 | 50.9 | 64.8 | 78.6 | 93.8 | 108.9 | Inches. |
| " " " " No. 3.... | ........ | 27.5 | 32.8 | 39.1 | 50 | 62.2 | 73.9 | 86.7 | 99.2 | " |
| | 1 in 120 | 78 | 143 | 257 | 590 | 1200 | 2180 | 3700 | 5825 | Acres. |
| Calculated size, by formula No. 2.... | ........ | 26.9 | 32.9 | 40 | 52.7 | 66.8 | 81.5 | 97.2 | 113.1 | Inches. |
| " " " " No. 3.... | ........ | 28.5 | 34.2 | 40.8 | 52.4 | 64.8 | 77.5 | 90.9 | 104.1 | " |
| | 1 in 80 | 90 | 165 | 295 | 670 | 1385 | 2486 | 4225 | 6625 | Acres. |
| Calculated size, by formula No. 2.... | ........ | 26.3 | 32.2 | 39.1 | 51.4 | 65.5 | 79.6 | 95 | 110.4 | Inches. |
| " " " " No. 3.... | ........ | 28.6 | 34.3 | 40.8 | 52.2 | 65 | 77.4 | 90.8 | 103.9 | " |
| | 1 in 60 | 115 | 182 | 318 | 730 | 1500 | 2675 | 4550 | 7125 | Acres. |
| Calculated size, by formula No. 2.... | ........ | 27.2 | 31.7 | 38.2 | 50.4 | 64.2 | 77.7 | 92.8 | 107.8 | Inches. |
| " " " " No. 3.... | ........ | 29.9 | 34.3 | 40.6 | 52.1 | 64.7 | 76.9 | 90.2 | 103.2 | " |

The sewer diameters given by the formula are in both cases greater than those given by Mr. Roe for the same areas and inclinations of sewers. Mr. Bazalgette's formula keeps closer to Mr. Roe's dimensions than the formula given by me, and they both exceed more in the larger areas than in the small ones, which is so far safe.

For large sewers the dimensions indicated here, in the shape of diameters of circles would take in practice forms of greater strength and safety.

What has been said on this part of the subject though somewhat irrelevant to the matter immediately in hand, may be of use to others feeling their way in the same direction. Our experience in works of this kind, in the United States, has been heretofore limited and unsatisfactory. The works built in Brooklyn operate very successfully, but until some pains shall be taken to observe and record for a while, their conditions under heavy rain-falls, we shall not know whether the dimensions adopted are just right, or whether they could have been reduced with safety. To insure their not being too small, it followed almost necessarily, that in this case they should be somewhat too large. I know of but one instance in which, from accidental circumstances, a piece of pipe was laid in opposition to this policy, which during a severe rain-storm proved to be too small for the service required of it. It occurs in the basin of the sewerage Map A, and I give below, as of interest in this connection a tabular comparison of the actual dimensions of the main sewer in this case, and the dimensions which the formula No. 1 (similar in its results to No. 2) would have given.

*Details of Main Sewer in Map A and length of its contributing Street Branches.*

| Number of Inclines. | Separate inclines of main sewer, Lengths. | Grade of these inclines. | Total lengths on Main Sewer, from its origin. | Total Lengths of entering street branches in rear of this point. | Aggregate lengths, main and its branches. | Maximum discharge for each point as per last column. | Diameter of Sewer for each incline. | | |
|---|---|---|---|---|---|---|---|---|---|
| | | | | | | | As built. | Calculated Size. | |
| | feet. | | feet. | feet. | feet. | cu. ft. per sec. | inches. | inches. | |
| 1 | 684 | 1 in 71 | 684 | .... | 684 | 2 | 12 | 9 | pipe. |
| 2 | 300 | 1 in 57.2 | 984 | 410 | 1,394 | 4.2 | 15 | 12 | pipe. |
| 3 | 1,541 | 1 in 87.7 | 2,525 | 3,580 | 6,105 | 18.3 | 16×24 | 19.8 | brick. |
| 4 | 1,096 | 1 in 154 | 3,621 | 3,980 | 7,601 | 22.8 | 18 | 23.4 | pipe. |
| 5 | 466 | 1 in 38.5 | 4,087 | 4,190 | 8,277 | 24.8 | 18 | 19.1 | pipe. |
| 6 | 450 | 1 in 26.8 | 4,537 | 4,190 | 8,727 | 26.2 | 18 | 18.3 | pipe. |
| 7 | 250 | 1 in 62.8 | 4,787 | 17,620 | 22,407 | 67.2 | 36 | 29 | brick. |
| 8 | 200 | 1 in 76.6 | 4,987 | 19,080 | 24,067 | 72 | 36 | 30.5 | brick. |
| 1 | 2 | 3 | 4 | 5 | 6 | 7 | 8 | 9 | |

A disappointment in the receipt of pipes led to the building of the third incline of the above sewer (16 × 24) of brick. The discharge in column 9, is obtained from the multiplication of the aggregate lengths of all contributing street sewers, main and branches in rear of the given point, by 0.003, or by 0.3 for 100 feet in length of street as elsewhere explained. The difficulty in this sewer occurred between points 3 and 4 where a pipe sewer is laid of 18 inches in diameter. The formula requires 23 inches at this point, and would have prescribed a larger size for the whole length of the 18-inch pipe. The other portions of the sewer are of larger size than the formula would require, and this is the case with all the sewers of the city except at some of the outlets, where the size determined on or built corresponds sometimes very closely with the size by this calculation.

The pipes generally used in the Brooklyn sewerage works have been strong glazed earthenware pipes of twelve, fifteen, and eighteen inches diameter. No pipes of less than twelve inches diameter have been used for a street sewer, although the entering sewers from dwellings do not exceed six inches in diameter. The pipes are of three feet in length each, and have been laid with sleeves, by which a better joint can be secured than with socket pipes. The necessary branch pipes are invariably laid as the sewer is built, to obviate the necessity of breaking into it afterwards. The branch opening is covered until wanted.

For the flow of sewerage the glazed pipe is far superior to the brick sewer, and now that we have had considerable experience of them, it is surprising that we do not venture, as in England, upon sizes exceeding eighteen inches diameter, up to twenty-four inches, at least. The pipe sewer has a joint every thirty-six inches; the brick sewer has a joint every eight and a half inches longitudinally, and every two and a half inches transversely. These mortar joints do not always remain tight. Whether from the use occasionally of ill-prepared mortar, or by some action of the sewerage matter on the mortar, they frequently leak. If the leakage is inwards it will assist the flow; but if the leakage is outwards, as will generally be the case in our city, it deprives the heavy matter of the sewerage proper of its fluid motor, and must lead to some settlement and deposit on the bottom, as a matter of course. The leakage is objectionable, besides, as containing in some measure that impurity of the subsoil which it was the object of sewerage works to correct. In all this the advantage is entirely with the pipe sewer, and it is for that reason worth more to the city than a brick pipe of the same size, and a larger price, if need be, might well be paid for it. The cheapest mode of building the sewers is not, in my opinion, the best, while a more perfect mode is within reach at but little extra expense. The roughness of all brick sewers, however well built, as compared with proper pipe sewers, and the foothold they give to rats, might well lead us to dispense with their use except for sizes where the more perfect mode becomes impracticable or unsafe.

The value of the sewers of either description depends essentially on the correctness of their lines, and on such faithful construction as will maintain those lines intact without settlement. If the sewer becomes leaky it will be difficult to avoid settlement or distortion of some kind. Great care is, therefore, sought to be taken in their construction, and in the stamping and packing of the earth around and over them after they are built. The leading sizes and forms of the brick sewers are given in Plate 55, to which the reader is referred. Man-holes are built along the line of the sewers at distances of 100 to 150 feet apart, their shape and size is given in Plate 56. They are covered at the surface of the street with cast-iron covers, in which there are holes for ventilation. This is the only means provided for the ventilation of the sewers, although the leaders from the roofs of the neighboring dwellings, where these are carried into the house-drain, must, to some extent, assist in ventilation.

At the corners of the streets are placed the gullies by which the rain-water reaches the sewers, these gullies are trapped, as shown in Plate 56, to receive the heaviest portions of the street washings, which settle in the trap, and until the trap is full cannot reach the sewer. If the traps are neglected, the waters of rain-storms carry all the washings of the neglected streets into the sewer, at the risk of producing obstructions and of causing considerable inconvenience and expense. If the cleaning of the streets receives reasonable attention from the authorities, the sewers may be expected to perform the important service which belongs to them with uniform satisfaction.

A large and important sewer is now being built to drain a sewerage district of two thousand three hundred acres. The form and dimensions of this sewer, towards its outlet in the East river, are given on Plate 57. All the Brooklyn main sewers deliver into tide-water, and where advantage can be taken of piers at the point of discharge, they are carried out to the end of the pier, that the sewerage waters may reach the tidal currents and be carried away by them. For the sewers draining small areas, a cast-iron pipe has been used to carry the sewer to the end of the pier. Where the sewers drain large areas, as in the case of the great sewer last mentioned, this is not practicable, and the arrangement for the terminus requires much consideration.

The rise and fall of the tides average five feet. The bottom of the outlet of the sewer is generally placed about a foot above low water-mark. The tide rises in the sewer, and the sewerage water proper either meets still water towards the outlet or an opposing current for probably eight hours out of the twelve. When the tide is at its lowest the sewerage water delivers into space under circumstances favorable to a very free discharge. This can happen, however, during not more, we shall say, than one third of the twenty-four hours. During the other two thirds the sewer is in effect obstructed towards its mouth, and a periodical settlement of all the heavier matters of sewerage must necessarily have place there. The size of outlet which shall best meet this condition of things requires special study, and can only be understood for each locality after some experience. Sluices or tide-gates at such points, and daily flushing, will be eventually necessary to prevent the accumulations of offensive matter, which, under a dense population, would otherwise have place there.

To determine the limits of these sewerage basins, the whole city was carefully levelled and a map prepared, showing the height above tide of every street corner and occasionally of intermediate points. The topographical features of the ground generally indicated the outlines of the basins, but these were sometimes extended or modified to meet other requirements. The Engineer has always striven to carry the outlet sewers into the tidal currents, and to avoid delivering into eddies or docks where the sewerage matter would be retained, and would eventually become offensive.

At this date the amount of brick and pipe sewers built, in accordance with this general system of sewerage for the city of Brooklyn, is as follows:

| | |
|---|---|
| Number of miles of 12 inch pipe sewers built | 53.48 |
| " " 15 " " " | 19.66 |
| " " 18 " " " | 11.20 |
| " " 24 " " " | 0.23 |
| " " 24 " brick " | 6.11 |
| " " 36 " " " | 3.32 |
| " " 42 " " " | 0.13 |
| " " 48 " " " | 1.64 |
| " " 54 " " " | 0.25 |
| " " 60 " " " | 1.15 |
| " " 72 " " " | 0.47 |
| " " 102 " " " | 0.08 |
| " " 108 " " " | 0.04 |
| Total | 97.76 |

The reader is referred to the sewerage plan for the position and size of each sewer, as well as for the relative positions of the sewerage basins.

JAMES P. KIRKWOOD.

BROOKLYN, *January* 24, 1865.

# APPENDIX.

## REPORTS BY MECHANICAL ENGINEERS

ON TRIALS OF DUTY

## MADE IN 1857 AND 1859,

UPON THE

## BROOKLYN, HARTFORD, BELLEVILLE, & CAMBRIDGE

PUMPING ENGINES.

# PREFATORY.

## FEBRUARY, 1860.

THE reports which follow are printed by order of the Board of Water Commissioners, both as presenting information of special interest to their fellow-citizens and as contributions, which they have not felt at liberty to withhold, toward the elucidation of a subject rendered frequently obscure to the general reader, more by the unreasonable claims and assertions of inventors and experts than by the inherent difficulties of the subject itself.

The causes which led to the experimental examination of other pumping engines than our own may not be familiar to all, and will therefore be briefly stated.

The contract of H. S. Welles & Co., for the entire works, includes the construction of two steam-pumping engines, each of capacity to deliver ten millions of New York gallons in sixteen hours (= 22.236 cubic feet per second), under a lift by construction of 162 feet and through a 36-inch main, 3,400 feet in length.

The contract describes and specifies a single-acting beam-engine, of the kind usually known as the Cornish engine.

The capacity of engine prescribed is unusually large. The diameter of cylinder required would exceed that of any other pumping engine here or in Europe, except the Leeghwater engine. The horse-power exceeds any of these, the Leeghwater included. Such an engine of the Cornish variety, with a stroke of ten feet, would have required a cylinder, it is estimated, of 126 inches diameter, and a pump or plunger of 54 inches diameter.

Four of the most prominent machine establishments of the country (the Novelty Works, the Hartford Works, the West Point Foundry, and Morris & Co., of Philadelphia) estimated severally the diameter of the steam cylinder at 111, 120, 136, and 160 inches ; three of them advised against the construction of so large a machine, and advised its division into at least two parts, making four engines instead of two. On the other hand, an engine of novel construction, combining ingeniously certain old inventions with much that was new, had been built, and was in use at Hartford, Connecticut, supplying the city with water. This engine was confidently asserted to be superior both in action and economy to the Cornish engine. It had acquired considerable reputation and notoriety among engineers, and those interested in it presented strong claims to be permitted to compete for a machine of the capacity required for the Brooklyn works.

The above considerations, and others of similar import, induced me to advise that the engineering talent of the country should be invited to compete for the construction of the required machine, under certain restrictions, which will be best understood by reference to the notice to machinists then circulated from this office, of which the following is a copy:

## "NASSAU WATER COMPANY, BROOKLYN, NEW YORK.

### MEMORANDUM TO MACHINISTS

*Designing to propose for the Construction of the two Pumping Engines required on the above Water Works.*

"The Engines are proposed to be of the Cornish class; but any other plan of engine may be submitted, and will be considered, that will perform the same duty as efficiently, and as economically.

"Each engine must be competent to perform, with ease, the following duty:

"To deliver into a reservoir, situated 170 feet above the pump-well, ten millions New York gallons daily, in sixteen hours, through a 36-inch main, 5,800 feet in length.

"The New York gallon of water weighs eight pounds.

"To be each connected with not less than three boilers, by a sufficient steam-pipe, &c.

"To be provided with two Worthington steam-pumps No. 5, properly connected with every boiler.

"To be provided each with an air chamber, built of boiler-plate iron, of not less than 4,100 cubic feet capacity.

"The person proposing to build the engine, may specify a different-sized air chamber, or a stand pipe, if he consider the last essential to the successful working of his engine, giving his reasons accordingly.

"The duty of each engine, with their boilers, to be further tested, by the standard of not less than 600,000 pounds of water, raised one foot high, with one pound of coal.

"The iron and composition work of each engine, all the materials used, and the workmanship and finish, to be in all respects thorough, and the first of its class; the dimensions, character, power, and action, of the separate parts and conveniences, to be of the most approved description in use, on the class of engines required; and these and everything connected with the engines, to be subject to the inspection and approval of the engineer of the said Nassau Water Company.

"The builder of the engine to furnish everything necessary to the placing it in position for work, and in the best working order, except the stonework of its foundations, which will be prepared for it, to suit the plan adopted.

"The builder to deliver and put up the engine in its proper place, connecting it with the 36-inch main aforesaid.

"Plans in detail of the engine proposed, shall accompany each offer, with a description of the parts, material, finish, conveniences, &c., understood and included in the offer.

"Offers and plans may be addressed, until the first of October next, to H. S. Welles & Co., No. 4 Wall street, New York, or to J. P. Kirkwood, Engineer, Nassau Water Co., Brooklyn, New York, of whom further information may be obtained.

"One of the engines must be put up, and in all respects ready for the required duty, on or before the 1st of January, 1858, and the second, on or before the 1st of January, 1859."

Twelve offers for furnishing the required power followed this notice, founded on varieties of single-acting and double-acting beam-engines, and of some direct-action engines. The different plans of these were carefully examined and analyzed at my request, by Mr. Samuel McElroy, who reported in favor of the double-acting beam-engine of Mr. William Wright of Hartford, as possessing certain advantages over all the others. This is the plan of engine since constructed, except that the steam cylinder has been built of 90 inches diameter instead of 100 inches as first proposed.

The notice to machinists above referred to, contains the following paragraph:

"The engines are proposed to be of the Cornish class; but any other plan of engine may be submitted, and will be considered, that will perform the same duty as effectually and as economically."

To ascertain whether an engine upon the model of the Hartford machine (which still claimed for itself the precedence), would perform as well and economically as the Cornish engine, it was necessary to make an experimental trial of the two. The Bellville engine, as the best specimen of a Cornish engine in the United States, was selected to represent its class; and Mr. Brevoort and myself were authorized to employ Mr. C. W. Copeland and Mr. W. E. Worthen, both well known mechanical engineers, as experts, to make the experiments. Their report is the first in order of these papers.

Mr. Worthington's offer to build a pumping engine after the model of the engine built by him for the Cambridge water works, led, for the same reasons, to an experimental trial of that engine, by Messrs. Morris and McElroy, whose report is the second in order. Mr. McElroy having afterwards made a separate statement of that experiment, founded upon what he considered a more correct mode than the ordinary one, of estimating the "duty" and showing a less favorable result than appeared in his joint report with Mr. Morris, a second trial of the Cambridge engine was directed to be made by a new Board, consisting of Mr. Frederick Graff and Mr. Erastus W. Smith; their report is third in order.

The last report of the series gives the result of a second trial of the Hartford engine, made at the desire of Mr. Wright. The second trial was conducted, with Mr. Wright's consent, by Mr. Graff and Mr. Worthen, and gives the results by two modes of estimating the fuel consumed; experiment No. 1 following what Mr. Wright considered the best mode, and experiment No. 2 being in accordance with that pursued by Messrs. Copeland and Worthen.

The reader is reminded that the object of these trials was to ascertain the working merits of the engines tried, as compared with the class of engine defined in the contract. In no other way could the assertions of the parties desiring to build engines upon these models, have been satisfactorily confuted or sustained.

The results furnish a sufficient basis upon which to form a fair judgment, not, however, by a simple comparison of these results, but after some allowance made for difference in size, it being a well-established fact that an engine of two-hundred-horse power will perform double the work of an engine of one-hundred-horse power, at less than double the expenditure of fuel; in other words, will show a higher "duty."

The term "duty" has been applied to a conventional measure of the economical value of an engine, as regards fuel simply. The true economic value must include the elements of first cost, attendance, maintenance, and repairs, as well as fuel. A high duty might therefore be shown by an engine in other respects deficient, and it is only when all the characteristics alluded to work profitably together that the machine takes high rank.

The report of Messrs. Copeland and Worthen gives various modes of estimating the "duty," the most usual of which seems to be that form of estimating it which obtains at the mines, called in that report the Cornish mode. This gives the engine the full stroke of the pump, without deduction for leakage and other losses there, and, on the other hand, gives the engine no credit for the friction of the water in the pipes. At the mines the one may about balance the other, and where there are so many engines to be reported, a more correct mode might demand a nicety of observation which would defeat the system of monthly reports.

In making "duty" trials of pumping-engines of city water works, a different course is necessary to admit of a fair comparison of results.

If the friction of the main is not credited to the engine, an engine which works into a small main will compare unfavorably with an engine, in all respects alike, working into a large main. The results in this case would be false, as tests of the efficiency of the machines, and to make them true it is necessary to include the work due to water friction, whether that arises from differences in length, or in size of pipe, or both combined. In the same manner the engine should have credit only for the water delivered by the pump, not for the length of stroke. In some pumps the loss of action does not exceed two per cent., while in others it reaches ten per cent. Where the engine pumps into a reservoir the actual delivery of the pumps per stroke can be ascertained very closely; where this is not the case other modes of appreciating the losses can be applied.

In the experiments lately made on the Brooklyn engine my desire was that the length of stroke for the pump should be estimated from the actual delivery of water into the reservoir, and not by the measured stroke, but an obscurity of language in the paper agreed upon between the parties, as to the mode of taking the duty, gave the referees a different impression, and led them to credit the engine with the measured stroke. The losses at the pump in this case were found to be but one and four fifths per cent., and the "duty," with this correction, would still have exceeded the contract provision.

Messrs. Copeland and Worthen supposed that the ordinary duty of the engines experimented on by them, might differ ten per cent. from their performance on a short trial, made with great care, under the eyes of their respective builders. Mr. Wicksteed's results show a larger loss; caused mainly, however, by the inferior quality of coal used during the eleven and a half years referred to. Using the same quality of coal as in the experiment, the difference, as shown by Mr. Wicksteed's statement, would not have exceeded five per cent.

JAMES P. KIRKWOOD.

ENGINEER'S OFFICE, BROOKLYN, *February* 20, 1860.

# REPORTS.

*Report of Messrs. Copeland and Worthen on the Belleville and Hartford Pumping Engines.*

NEW YORK, *January* 30, 1857.

JAMES P. KIRKWOOD, *Engineer*, J. CARSON BREVOORT, *Secretary*,
*Nassau Water Company :*

GENTLEMEN : In accordance with your note of November 11th, ult., we have made trials of the actual duty performed by the pumping engines at Hartford and Belleville, and herewith submit the following report :

Previous to commencing the trials, due notice of our intention was communicated to the engineers of the different works, to the builders of the engines, and to the patentee of the Hartford pumps, with the suggestion that the machinery should be put in an order satisfactory to themselves, which suggestion was carried out at Hartford by Mr. Wright for Woodruff and Beach ; and by Mr. Worden, at Belleville, for the Cold Spring Works. These parties were present at the trial of their respective engines during the whole course of the experiments.

The experiments were conducted as far as possible in the same order and manner at both places. The levels were taken by the same engineer, Mr. McElroy ; the same assistants were employed as observers, and in similar positions, and the same fireman was employed in firing. It was also the intention to have used exactly similar coal, but owing to the ice in the river the coal shipped to Hartford did not arrive in time ; its place was therefore supplied by purchase from a yard at Hartford—a similar quality of Cumberland from George's Creek, but not from the identical mine with that forwarded and used at Belleville ; it was somewhat larger, and in rather better condition than the latter. All requisite checks were applied to the observations, and although drawings and dimensions were furnished by the builders of the engines, yet all important parts were subject to measurement.

### DESCRIPTION OF WORKS.

At the Hartford Works the boiler is a drop-flue boiler and very well clothed. The chimney is very high, and the draft good, but the damper is generally open but very little.

The back end of the boiler and all steam connections were felted.

The engine is a double acting, condensing crank and beam engine, with a single cylinder of $32\frac{3}{8}$-inches diameter, five feet stroke, with an adjustable cut-off of Wright's patent. The injection water is taken from the well. There is a heavy fly-wheel of twenty-two feet diameter on the crank-shaft, and power is communicated by a pinion of twenty-seven teeth on the end of this shaft, gearing into a spur-wheel on either side of eighty teeth, and on each spur-wheel shaft are two cams, each giving motion to a set of pumps by means of bell-cranks. Each set of pumps consists of two pistons or boxes in one chamber or cylinder, one above the other, the piston-rod of the upper one being a tube, and the piston-rod of the lower box passing through it; the valves are butterfly-valves, hinged in the middle; and each piston commences its stroke before that of the other has been completed; each commencing its stroke slowly, and increasing in a short space to its uniform velocity, and at the end decreasing for like distance till it stops. Thus, while the lower box is rising the upper is descending the water passing up through the valves of the upper box; but just before the lower box has completed its up-stroke, the upper box has completed its down stroke, and commences to rise, the lower decreasing in velocity and the upper increasing. And *vice versa*, during the rise of the upper box the lower one descends, and commences to rise in time to relieve the upper box at the end of its stroke. Thus the stroke of one box laps on to that of the other, and of the absolute movement or stroke of $17\frac{7}{8}$ inches of each box we have considered but $16\frac{1}{8}$ inches as effective.

### THE MAINS.

The water is brought from the river to the supply-well, some fifteen feet outside the building, by a pipe two feet in diameter. The pumps are arranged in plan in the form of a rectangle; a suction-pipe is brought up on each side, and the water is delivered through the pumps into a central rising main, between the pumps, which is sixteen inches in interior diameter, and is provided with a check-valve at the bottom and a small air-chamber. In its route to the Reservoir it is laid through the streets most convenient, following their courses and in general their grades, and is introduced into the bottom of the Reservoir. The length of the rising main is about six thousand nine hundred feet.

The Reservoir is nearly rectangular in form, with earth banks and an interior puddle-bank of clay: the inside slope is protected by rubble.

### BELLEVILLE.

The boilers, four in number, are of the usual Cornish pattern, well covered; the chimney is high, and draft good, and, as at Hartford, is controlled by a damper. But three boilers were used during our experiments.

The steam connections were all felted.

The engine is a single acting Cornish engine of $80\frac{3}{16}$ inches diameter of cylinder, and a maximum stroke of eleven feet. The cylinder is protected by a steam jacket; the injection water is taken from the rising main.

The plunger of the pump is 34¾ inches diameter, and the stroke the same as that of the engine.

The water is introduced from the river by a culvert some three hundred and seventy-five feet in length, into the supply-well, which is beneath the engine-house.

The rising main is thirty-six inches interior diameter, with a check valve, some ten feet vertical height above the top of the pump chamber. It enters the Reservoir level with the bottom, is turned up with a curve, and discharges with an overflow three feet below the top of the bank. About three hundred feet north of the Reservoir there is a summit from which, on one side the main descends towards the engine-house, and on the other it falls about eighteen inches to the level of the bottom of the Reservoir.

The whole length of the pipe is about two thousand three hundred feet.

The Reservoir has earthen banks, with an interior puddle wall, and lined with brick four inches in thickness laid on a bed of cement plaster.

## MODE OF CONDUCTING EXPERIMENTS.

### HARTFORD.

Previous to the commencement of the trial the boiler was fired for a considerable length of time, sufficient to thoroughly heat the brickwork and the stack; the fire was brought into a fair average condition, the height of water in the boiler to the proper line, and the steam to its usual working pressure; these were maintained as nearly constant as possible, during the course of the experiment, and left in the same condition at its close.

The height of the water in the boiler was taken by a glass-gauge.

The coal was weighed, in lots generally of two hundred pounds, on scales tested and certified to, by the sealer of weights and measures of the city.

To measure the quantity of water evaporated by the boiler, a tank was provided into which the feed water was delivered and measured by accurate gauges attached to the tank, and temperature noted; this was then supplied to the boiler by a Worthington steam pump. The steam from this pump was passed through a coil in the tank, condensed, and the water discharged outside, in this way abstracting almost entirely the heat used to drive the pumps, and restoring it to the boiler.

In estimating the evaporation by volumes, this steam should be taken into consideration, but as the loss in this way is trifling, and unimportant to the comparative values of the results obtained from the two engines, it has been neglected.

At the engine, indicator-cards were taken every half hour, in the first experiment, at the top only, but in the second experiment at both ends of the cylinder. At each half hour, also, the number of strokes both of engine and pumps, the pressure on the rising main, the vacuum on the suction, the height of water in the supply-well, the height of water and pressure of steam in the boiler, were noted. At the Reservoir the height of water was taken every fifteen minutes. Five observers besides ourselves were employed; their watches were brought to one standard, and the observations were simultaneous.

It was the intention to take indicator cards from the rising main, but the oscillations of the piston of the instrument were of so little range as to furnish but an indifferent card; an Ashcroft pressure gauge was therefore substituted from which the readings were taken. At a subsequent date indicator-cards were taken from the pump chambers; the water in the Reservoir and supply-well being of about the same height as during the experiment.

## BELLEVILLE.

The same course was followed as at Hartford in maintaining a constant fire, and pressure and height of water in the boilers during the course of the experiment and to its close. The scales were tested and sealed by the officer from Newark. Identical arrangements were also adopted for measuring the evaporating power of the boiler.

The water condensed in the steam jacket is returned directly to the boiler, and this amount we had no means of measuring.

Indicator-cards were taken from the rising main and from top of cylinder; the vacuum beneath the piston, the counter, the length of stroke of the pump, the vacuum in the suction, the height of water and pressure of steam in the boilers, and height of water in the supply-well, were noted every half-hour. The height of water in the reservoir was taken every fifteen minutes. The indicator-cards from the rising main were taken by a continuous movement connected with clockwork; and at a subsequent date indicator-cards were taken, as at Hartford, directly from the pump-chamber, and also vacuum cards from beneath the piston of the engine, to ascertain the action of the lower valve.

During the second experiment at Belleville, the coal being exposed to the rain, became wet; at 10.30 A. M., 319 pounds of coal were weighed and placed in the fire-box of the boiler not in use. At the conclusion of the experiment, 2.30 P. M., it was drawn, and although not entirely freed from moisture, the same coal weighed but 300 pounds; 6.3 per cent. have therefore been added to the evaporation and duty. The injection-water at Belleville was taken from the main, but having no means of estimating it accurately, it is not included in the duty as given in the above table.

# TABLE I.—Boilers.

| | No. of Experiment. | Duration of Experiment. | Number and kind of Boiler. | Dimensions of Boilers. | Total Heating Surface. | Total Grate Surface. | Average Pressure of Steam on Boiler, per square inch. | Total Coal consumed during Experiment. | Coal consumed per square foot of Grate per hour. | BY MEASURE. Total Evaporation. | BY MEASURE. Evaporation per pound of Coal. | BY VOLUMES. Equivalent Evaporation per lb. of Coal from standard temp. 100°. | BY VOLUMES. Total Evaporation from temp. 32°. | BY VOLUMES. Evaporation per lb. of Coal from temp. 32°. | BY VOLUMES. Equivalent Evaporation per lb. of Coal from standard temp. 100°. |
|---|---|---|---|---|---|---|---|---|---|---|---|---|---|---|---|
| | | H. M. | | | sq. feet. | sq. ft. | lbs. | lbs. | lbs. | lbs. | lbs. | lbs. | lbs. | lbs. | lbs. |
| HARTFORD. | | | | | | | | | | | | | | | |
| Dec. 5th...... | 1st. | 2.30 | One Drop Flue. | 7 feet 6 in. diameter, 22 feet 8 in. long. | 1112.49 | 23.125 | 29.5 | 400 | 6.918 | From temp. 100.°5 4007.48 | From temp. 100.°5 10.019 | 10.015 | 2297.664 | 5.7442 | 6.0946 |
| " 5th & 6th. | 2d. | 8.30 | Do. do. | Do. do. | Do. | Do. | 26.14 | 1117 | 5.683 | From temp. 84.°5 12074.48 | From temp. 84.°5 10.810 | 10.961 | 8841.781 | 7.915 | 8.398 |
| BELLEVILLE. | | | | | | | | | | | | | | | |
| Dec. 19th..... | 1st. | 6.00 | Four Cornish boilers. Three only in use. | 7 feet diameter, 34 feet long. | of 3 boilers. 2450.25 | 60.666 | 26.06 | 1806 | 4.961 | From temp. 91.° 17159.037 | From temp. 91° 9.700 | 9.778 | 12751.072 | 7.060 | 7.491 |
| " 20th..... | 2d. | 12.00 | Do. | Do. | Do. | Do. | 26.24 | 3524 | 4.839 | From temp. 92.°3 34590.624 | From temp. 92°3 9.816 | 9.884 | 24867.465 | 7.057 | 7.487 |
| " 20th..... | 2d. | ..... | .................. | .................. | .......... | ...... | ..... | .... | ..... | ............... | ................. | 10.507 | ......... | ...... | 7.960 |
| | 1 | 2 | 3 | 4 | 5 | 6 | 7 | 8 | 9 | 10 | 11 | 12 | 13 | 14 | 15 |

# TABLE II.—Engines.

| | No. of Experiment. | Duration of Experiment. | Kind of Engine. | Diameter of Cylinder. | Actual length of Stroke. | Average Total Effective Pressure. | Average Total Press at End Stroke. | Number of Pumps. | Diameter of Pumps. | Diameter of Cylinder Piston Rod. | Diameter of Pump Piston Rod. | Stroke of Pumps. | Average Number of Strokes per Minute. | Average height Water is lifted. | Duties, or Pounds lifted One Foot high with One Hundred Pounds Coal. | | | | |
|---|---|---|---|---|---|---|---|---|---|---|---|---|---|---|---|---|---|---|---|
| | | | | | | | | | | | | | | | Power exerted upon Piston. | Work done at Pumps. | Work done at Rising Main. | Effective Duty, or Weight of Water delivered at Reservoir. | Duty according to Cornish Rule. |
| HARTFORD: | | H. M. | | ins. | feet. | lbs. per sq. inch. | lbs. per sq. inch. | | inches. | inches. | inches. | inches. | | feet. | | | | | |
| Dec. 5th ..... | 1st. | 2.30 | Beam and Crank Condensing Engine. | $32\frac{3}{8}$ | 5 | 9.002 | 3.9 | 8 | 4 of $19\frac{3}{8}$ and 4 of $18\frac{13}{16}$. | $3\frac{1}{2}$ | $4\frac{3}{8}$ and $3\frac{1}{8}$ | $16\frac{1}{8}$ | 8.13 | 119.063 | 67,171,604 | .......... | 55,068,516 | 44,200,903 | 51,865,914 |
| " 5th & 6th. | 2d.. | 8.30* | | $32\frac{3}{8}$ | 5 | 8.373 | 4.75 | 8 | | $3\frac{1}{2}$ | | $16\frac{1}{8}$ | 8.088 | 118.159 | 71,899,664 | .......... | 62,661,715 | 50,257,707 | 59,412,793 |
| BELLEVILLE: | | | | | | | | | | | | | | | | | | | |
| Dec. 19th.... | 1st. | 6.00 | Cornish. | $80\frac{3}{16}$ | 10.94 | 16.630 | 7.629 | 1 | $34\frac{3}{4}$ | $7\frac{3}{4}$ | Plunger Pump. | 10.94 | 4.805 | 159.824 | 87,832,671 | 72,115,396 | 71,801,323 | 62,515,827 | 68,804,844 |
| " 20th. ... | 2d.. | 12.00 | Do. | $80\frac{3}{16}$ | 10.906 | 16.641 | 7.937 | 1 | $34\frac{3}{4}$ | $7\frac{3}{4}$ | | 10.906 | 4.535 | 159.087 | 84,743,713 | 69,533,221 | 68,563,409 | 60,731,719 | 66,035,507 |
| " " ..... | 2d.. | ...... | ........ | .... | ...... | ........ | ...... | .. | ........ | ...... | | ...... | ...... | ....... | 90,082,567 | 73,913,814 | 72,822,903 | 64,557,817 | 70,195,744 |
| | 1. | 2. | 3. | 4. | 5. | 6. | 7. | 8. | 9. | 10. | 11. | 12. | 13. | 14. | 15. | 16. | 17. | 18. | 19. |

* During the course of this Experiment the Engine was stopped for twenty-one minutes, making the actual running time eight hours and nine minutes.

# TABLE III.

*Comparison of Capacity of the Pumps per Stroke, with the Absolute Discharge, and the Height as indicated by Gauges, with the Height as determined by the Level and by Observation.*

| NAME. | Number of Experiment. | CAPACITY OF STROKE. | | HEIGHT. | | |
|---|---|---|---|---|---|---|
| | | By Measure. | By Discharge. | By Level. | By Gauge at Pumps. | By Gauge on Mains. |
| | | Cubic feet. | Cubic feet. | feet. | feet. | feet. |
| Hartford.......................... | First .... | 20.532 | 19.665 | 119.063 | ...... | 139.651 |
| | Second .. | 20.532 | 19.259 | 118.159 | ...... | 137.678 |
| Belleville.......................... | First.... | 72.051 | 65.465 | 159.824 | 166.834 | 166.157 |
| | Second .. | 71.827 | 66.058 | 159.087 | ...... | 164.556 |

The pressures are converted into feet by multiplying by 2.3.

# TABLE IV.

*Duties of the Pumping Machinery reduced to a Common Standard of Evaporation of Ten Pounds of Water per Pound of Coal, from the Temperature of* 100 *Degrees Fahrenheit.*

| NAME. | WATER. | Power Exerted on Piston. | WORK DONE. | | Effective Duty at Reservoir. | Duty by Cornish Mode. |
|---|---|---|---|---|---|---|
| | | | At Pumps. | At Rising Main. | | |
| Hartford .............. | By measure. | 65,595,442 | ........ | 57,167,882 | 45,851,390 | 54,203,807 |
| Second experiment ..... | By volumes. | 85,614,865 | ......... | 74,495,969 | 59,844,852 | 70,746,360 |
| Belleville.............. | By measure. | 85,738,277 | 70,349,272 | 69,469,353 | 614,44,476 | 66,810,510 |
| Second experiment..... | By volumes. | 113,187,809 | 92,871,939 | 91,576,610 | 81,116,227 | 88,200,223 |

*Belleville, Second Experiment, with an Increase of* 6.3 *per cent. to compensate for the Wet Condition of the Coal. Standard of Evaporation—Ten Pounds of Water per Pound of Coal from the Temperature of* 100 *Degrees Fahrenheit.*

| NAME. | WATER. | Power Exerted on Piston. | WORK DONE. | | Effective Duty at Reservoir. | Duty by Cornish Mode. |
|---|---|---|---|---|---|---|
| | | | At Pumps. | At Rising Main. | | |
| Belleville.............. | By measure. | 91,139,788 | 74,781,276 | 73,845,922 | 65,315,478 | 71,019,572 |
| Second experiment ..... | By volumes. | 120,318,641 | 98,722,871 | 97,345,936 | 86,226,549 | 93,756,837 |

## EXPLANATION OF TABLES.

Table I. contains the evaporation of the boilers in the different experiments, together with the necessary description and dimensions.

*Columns* 1 *to* 9 inclusive, are sufficiently explained in the captions of the columns.

In reducing cubic feet of water to pounds—column 10—the weight of a cubic foot at different temperatures has been taken from table, p. 29, in Francis' "Lowell Hydraulic Experiments," or from an extension of the table when necessary.

The conversion of the evaporations from the observed temperature to that from a standard of 100 degrees, has been calculated from the English Admiralty formula:

"1212 degrees minus observed temperature, ÷ 1212 minus standard temperature, × actual evaporation," = Evaporation from standard temperature.

Thus, in 1st experiment at Hartford:

$$\frac{1212 - 100.5}{1212 - 100} \times 10.019 = 10.015.$$

To determine the evaporation by volumes, we have determined the net capacity of the cylinder, and the clearances, and the final pressure, and have divided the capacity by the volume corresponding to the final pressure. At Belleville, the equilibrium-valve closes before the completion of the up-stroke, and a portion of the steam is retained in the cylinder; this amount has also been determined from indicator cards and tables of volumes, and the amount deducted from the capacity as determined above.

The volumes corresponding to a given pressure have been determined from the Formula, page 75, and the Table, page 76, of Pambour's "Theory of the Steam Engine." The cubic feet of water thus found by the formula are referred to an initial temperature of 32°, and to bring the result to a comparison with that obtained by measure, we refer it by the English formula above given, to the same standard of one hundred degrees.

It will be observed that the evaporation by volumes per pound of coal, *columns* 14 and 15, in the first experiment at Hartford, is much less than the subsequent experiments; this is owing to the less quantity of steam used in the upper end of cylinder, from which alone cards were taken in the first experiment, and if addition were made in proportion for the excess of steam absolutely found in the up-stroke in the second experiment, the difference then in the evaporation would be trifling.

Table II. is the work done, or duties in pounds lifted one foot high with one hundred pounds of coal.

In *column* 12 of the Belleville experiments, it will be observed that the average length of stroke is not the same in both trials; the stroke of the Cornish engine is not constant and defined; it was therefore necessary to measure the stroke, which was done by a sliding gauge every half hour simultaneous with the other observations.

The heights, *column* 14, are determined by measures referred to a base of low water.

*Column* 15. The power exerted upon the piston was determined from the average pressure taken from the half-hourly indicator cards, from the area of the piston, the length of stroke, and the number of strokes per one hundred pounds of coal. At Belleville the average pressure above the equilibrium line is taken from the indicator cards from the top of the cylinder, and the pressure below the equilibrium line from the observed vacuum, and from indicator cards beneath the piston.

*Column* 16. Work done at pumps.

The pressure on the pump pistons is determined by an average of indicator cards which is multiplied by the area of the piston, or plunger, by the length of stroke, and by the number of strokes per one hundred pounds of coal. At Hartford, cards were taken above the upper box, and between the boxes, to which is added the vacuum of the suction pipe, and the pressure incident to the difference of level between vacuum gauge and indicator. The net area of each piston is multiplied by its average pressure and suction, and the average of the two multiplied by the length of stroke of each, by the number of pumps, and by the number of strokes per one hundred pounds of coal consumed.

*Column* 17. Work done at rising main, calculated from the average pressure on the rising main, as shown by a pressure gauge at Hartford, and by indicator cards at Belleville, to which is added the average vacuum on the suction, and the pressure due to the difference of level between the pressure and vacuum gauges, multiplied by the average net area of the pump pistons at Hartford, and by the area of the single plunger at Belleville, by the stroke, and by the number of strokes per one hundred pounds of coal.

*Column* 18. Effective duty, or weight of water delivered at the Reservoir. This is the absolute weight of water delivered at the Reservoir, obtained by multiplying the cubic contents as gauged at the Reservoir, by the heights obtained from level and from observation.

The heights taken at the reservoir were observed with great nicety—a scale with a sharp point in a glass tube being used at Hartford, and a hook-gauge (obtained from Mr. Francis of Lowell) at Belleville.

The temperature of water received in the pump-wells was noted, and found the same at both places—32° Fahrenheit. The weight of a cubic foot is reckoned at 62.375 pounds.

The leakage at the reservoirs was tested, and an addition has been made to the measures as gauged at Hartford, for a small leak, but no addition has been made at Belleville; on the contrary a small amount has been subtracted in the second experiment for the effect of rain.

At 3.30 A. M., a fall of rain commenced and continued throughout the trial; this rain-fall, at the level of the reservoir, was carefully gauged, and found to be .03 foot, which is, therefore, multiplied by the area of the reservoir, and the amount deducted from the contents gauged during the whole experiment.

*Column* 19 is the duty according to the Cornish rule; that is, reckoning each stroke of the pump at the full throw of its piston or plunger, and making no allowance for leaks through valves and pistons, or lap of strokes at Hartford, or for other losses of any description.

It will be observed, upon examination of Table I., that the evaporation per pound of coal from the boilers at Hartford and at Belleville differs to some extent, and as it is desirable to compare the efficiency of the machinery proper, independent of the economy of the different boilers, Table IV. was prepared, in which the absolute evaporation of the boilers is eliminated, and the relative efficiency of the machinery shown. The calculations are made both from the water evaporated by measure and by volumes.

The table is calculated thus: a standard boiler is assumed, the evaporation of which is ten pounds of water per pound of coal, and the evaporation in both cases is brought to the standard, which is in effect assuming that at both places the engines were supplied with steam by boilers evaporating ten pounds of water per pound of coal; this being the case, whatever difference there may be in duty is due to the machinery alone, by which the result is effected, and is shown by a simple inspection of the table.

As the second experiment in both trials is that which we consider the most reliable, Table IV. has been calculated for these experiments only.

Appended to Table IV. is a table showing the results obtained at Belleville, increased 6.3 per cent. for the wet condition of the coal.

With regard to the shorter duration of the experiments at Hartford, there was no separate distributing reservoir at that place, consequently, during the continuance of the experiment, the water must necessarily be shut off from the city, and of course the experiment tried in the night: the duration of the second experiment, eight and a half hours, being the longest time that the superintendent of the works thought it prudent to shut off the supply.

At Belleville, on the contrary, there was a large distributing reservoir, independent of the receiving reservoir, and the duration of the trial there was as long as we thought necessary.

That the modes of obtaining many of the results in the tables may be the better understood, we subjoin *fac-similes* of three indicator cards, from each set obtained, these three being about an average of the set of which they are a part.

*Figs.* A, A′, A″, show three cards taken from the upper end of cylinder at Hartford, and *Figs.* B, B′, B″, three cards taken at the same time (by a different indicator) from the lower end. These cards require no explanation.

*Figs.* C, C′, C″, show three cards taken from the upper of the two pumps at Hartford; it will be observed that these upper pumps (four of them) discharge into a common delivery pipe, in which is a delivery or check-valve, common to all the pumps, hence these cards represent very nearly the pressure within the rising main.

*Figs.* D, D′, D″, show three cards taken from the lower pumps at Hartford; it is necessary to a proper reading of these cards to recollect the peculiar operation of the upper and lower buckets of these pumps, as before described.

*Figs.* E, E′, E″, give three cards taken from the upper end of cylinder at Belleville.

To obtain a proper representation of the action of the steam within the cylinder, it is necessary to attach a card from the lower end of cylinder to one from the upper end, placing the atmospheric line of the one card upon that of the other.*

It should be observed, that, at the time the cards were taken from the lower end of the cylinder, the vacuum was not so perfect as during the experiments; hence the vacuum line is not so far below the equilibrium line as it would be on a correct card taken during the experiment.

*Figs.* G, G′, show two cards taken from the rising main at Belleville, as already described; it will be observed that the indicator returns partially, but not wholly, to the atmospheric line at the termination of each stroke; this effect is produced by a check-valve in the rising main, placed a few feet above where the indicator was attached; this valve of course relieved the indicator of the pressure of all that portion of the column of water above the valve, while the pump was making its up-stroke.

*Figs.* H, H′, are two cards taken from the pump on a day subsequent to that of the experiment; but, as at Belleville, the rising main is carried up through the water in the receiving reservoir to a point above its highest level, the head upon the pump thus always remaining the same; that portion of the card above the atmospheric line is a true representation of the resistance above the pump at all times.

As there was a rise and fall of the water in the supply-well corresponding to the rise and fall of the tide in the river, the vacuum of the card would vary; therefore in estimating the duty or work done at the pump, the actual level of the water below the pump was taken from observations during the experiments, instead of taking the vacuum as shown by the pump cards.

During the course of these experiments there was undoubtedly more care in firing, and more attention given to the condition of the machinery, than would be the case in ordinary daily work, and we are of the opinion that on this account a deduction of ten per cent. should be made from the result set forth by the tables to obtain a fair representation of the duty in their regular ordinary operation.

It was our intention in the report to have investigated fully the comparative value of the works at the two places in a commercial point of view, that is, taking into

* In the printed cuts a steam-card and a vacuum-card are given together.

consideration *first cost* as well as *duty* and *working expenses*. At our request, therefore, the superintendents at both places furnished us statements of the cost of the different machines, foundations, and structures; but on examination, we find it impossible to determine satisfactorily what proportion of the last two items is properly chargeable to the machinery, and what to the exigencies of the locality, and what to arrangements for extended supplies.

Thus at Hartford, the foundations are upon piles driven into quicksand, requiring extensive sheet piling; at Belleville, the foundations are upon a sandstone ledge.

At Hartford, as the river is subject to high freshets, the pump and air-chamber well and engine-room, are protected by a curb of solid masonry, carried above the high-water mark; no such precautions were necessary at Belleville, but foundations were there laid for an additional engine, looking to the necessity of a greater supply of water hereafter.

In view of these many diversities, we have compared the two engines commercially, merely with regard to their duties. Thus, by reference to the columns of work done at the rising main—Table II., experiment 2d—at Hartford, it will be seen that the duty is, in round numbers, sixty-two millions; at Belleville—2d experiment—with the addition for the condition of the coal, seventy-two millions.

If we estimate the value of coal delivered at the works at $6 per ton of 2,240 pounds, then the cost of one hundred millions of duty would be at

| | |
|---|---|
| Hartford | $0.432 |
| Belleville | $0.372 |

Or, in favor of Belleville....... .06 per one hundred millions of duty.

Each engine at Brooklyn is required to raise ten millions of gallons, of eight pounds each, 170 feet high, per day, or

| | | |
|---|---|---|
| | 13,600 | millions of duty. |
| Add for friction of pipe 15 per cent...... | 2,040 | " " " |
| | 15,640 | " " " |

Hence 156.40×.06=$9.38 per day in favor of the Belleville engine, as applied to *each* Brooklyn engine. The year's savings, allowing three hundred working days to the year, would be $2,814; or the interest at seven per cent. on $40,200, or, for the two engines, as required by the contract, $80,400.

At the time these experiments were entered upon by us, we had great doubt as to the results that would be obtained, and knowing there was a difference of opinion among engineers and others in regard to the relative efficiency of the two modes of elevating water for the supply of cities and towns, we endeavored to conduct the experiments with the utmost care, and with entire impartiality, having no other interest or object than to arrive correctly at the facts, and so report them. For this reason many checks were provided and independent observations taken, which ordinarily would be considered unnecessary; we believe no other course would have been just to ourselves or satisfactory to you.

In closing, we desire to express our obligations to Mr. McElroy, for the valuable assistance rendered by him, both in observations and calculations; to the Water Commissioners, engineers, and others connected with the works; particularly to Mr. Bissell, the Superintendent of the works at Hartford, and to Mr. Bailey, Superintendent of the Jersey City Water Works, for the promptness with which every facility was given, and assistance furnished for the prosecution of the experiments, and the more so, as it was evidently from a desire that no obstacle should be interposed to the attainment of the desired information.

Respectfully submitted.

CHARLES W. COPELAND,

W. E. WORTHEN.

# DIAGRAMS

*Accompanying the Report on Belleville and Hartford Engines.*

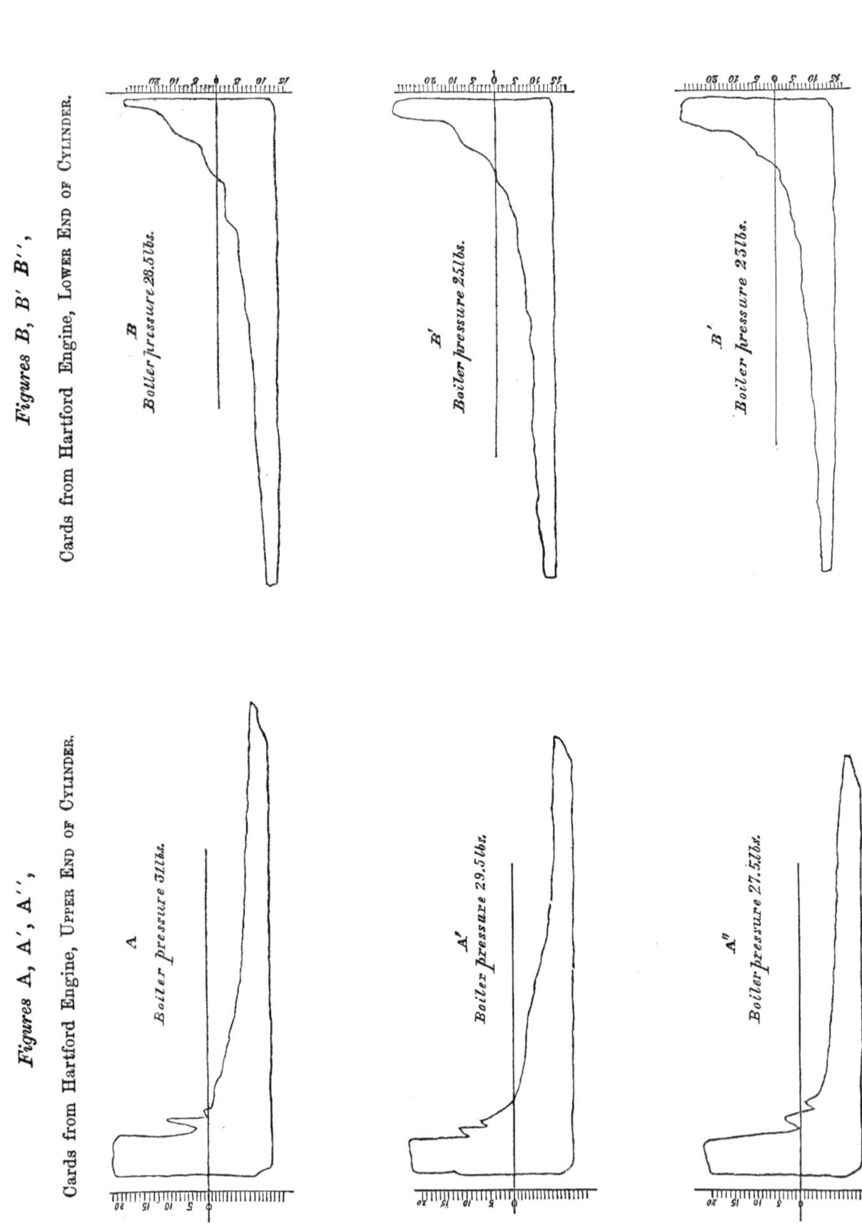


*Figures* B, B' B'',
Cards from Hartford Engine, Lower End of Cylinder.

*Figures* A, A', A'',
Cards from Hartford Engine, Upper End of Cylinder.

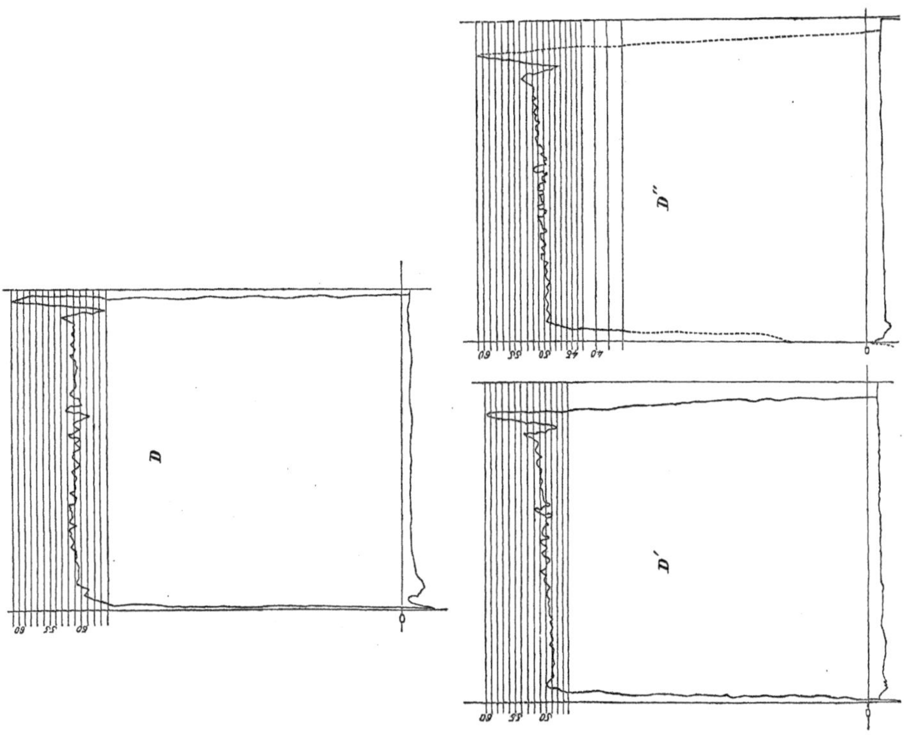

*Figures* D, D′, D″,
Cards from Hartford, Lower Pump.

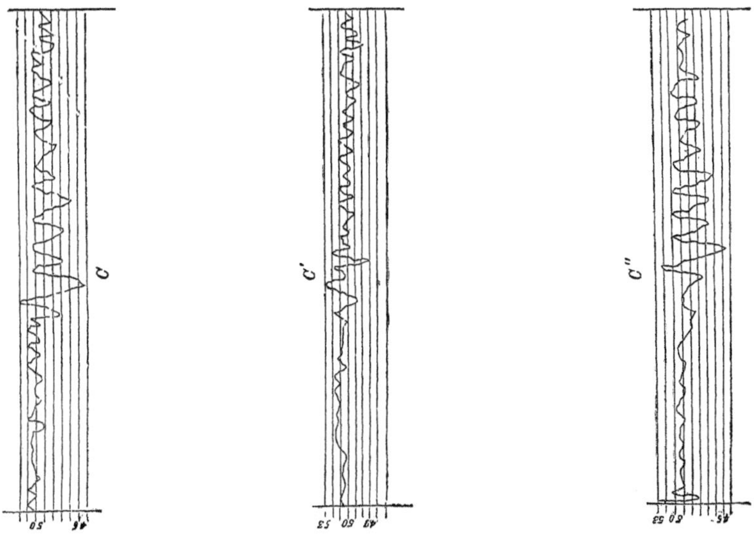

*Figures* C, C′, C″,
Cards from Hartford, Upper Pump.

*Figures* E, E′, E′′,

Engine Cards—Belleville, Steam and Vacuum.

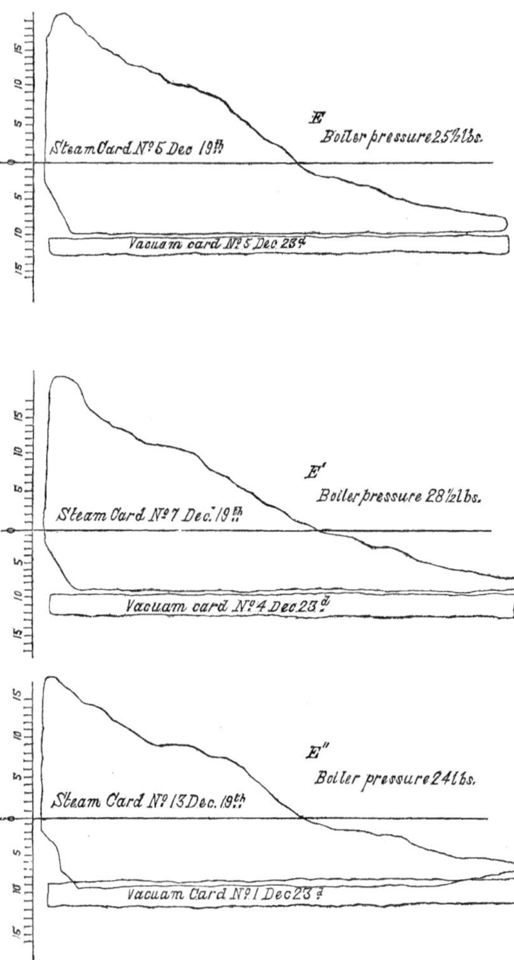

*Figures* G, G′, Cards at Rising Main—Belleville.

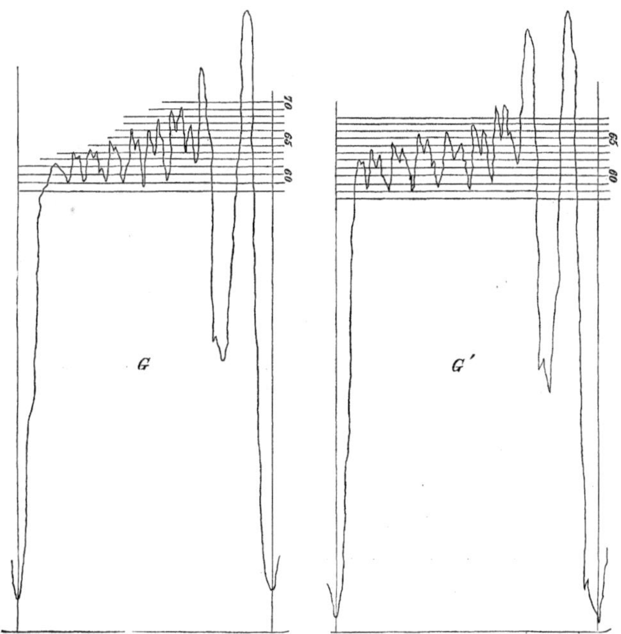


*Figures* H, H′, Pump Cards—Belleville.

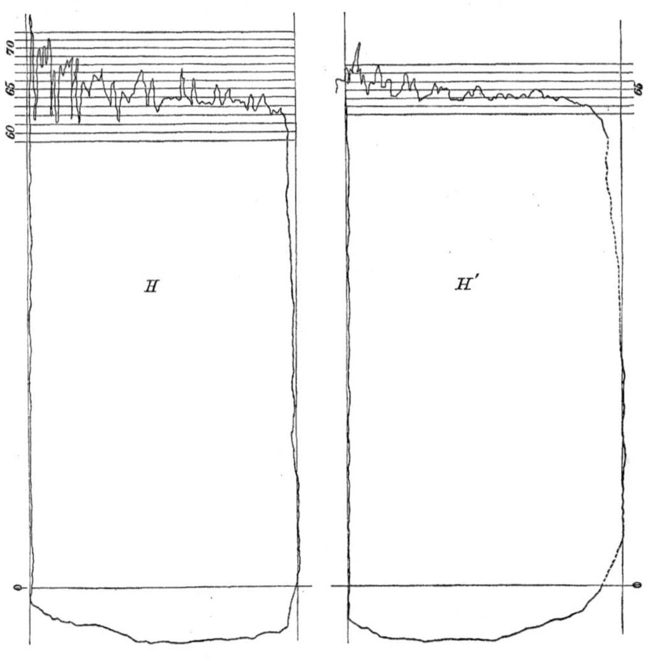

*Report by Messrs. Morris and McElroy, on the Cambridge Pumping Engine.*

BROOKLYN, *April* 9, 1857.

SIR : The undersigned, having made an examination of the Cambridge Pumping Engine, on the 30th ult., respectfully report :

That the engine and boiler used were taken warm, with clean furnaces, and fires started at 10 : 40 A. M., noting the wood and coal used, and that the engine was started at 11 : 38 A. M., with thirty-three pounds of steam in the boiler, and run until 12 : 8 P. M. (with the exception of four minutes stoppage), being run down from the total coal used, of which the last charge was put in at 9 : 40 P. M.

That, from the firing, and other notes taken on this occasion, we have decided to assume the condition of the boiler, at 12 : 30 M. and 9 : 30 P. M., as uniform, for the basis of a calculation of duty, which results as follows :

Steam.......................... 40 pounds pressure.
Water .......................... uniform level.
Fire.......................... uniform condition.
Coal consumed.................. eight equal charges, of 45 pounds each, put in at 12 : 30, 1 : 35, 2 : 37, 3 : 38, 4 : 45, 5 : 45, 7, and 8 : 20 P. M., being 360 pounds.
Counter (Stillman & Allen's) at 12 : 30........................ 123077
9 : 30........................ 133759
No. of double strokes............. 10682, or.. 21364 single strokes.
Net area of plunger.............. 156.6 sq. in. (rods deducted).
Length of stroke................. 2.181 ft.
Actual lift of water (taken from Company's Report) 72.29 ft.

CALCULATION OF CORNISH DUTY.

$$\frac{4.56 \text{ c. ft.} \times 62.38 \text{ lbs.} \times 10682 \text{ strokes} \times 72'.29}{3.6 \text{ bushels of } 100 \text{ lbs.}} = 61,015,243$$

CALCULATION OF DUTY AT DELIVERY.

150.6 sq. in. $\times$ 2′.181 $\times$ 36.15 lbs. $\times$ 5934.44 strokes per 100 lbs. coal = 70,463,750 ft. lbs.

The average pressure of the Ashcroft Gauge on the delivery was....... 31.9 lbs.
Average suction by indicator.......................................... 4.25

Total load.......................................... 36.15 lbs.

The calculations of duty are based on the notes to Table II., Report of Messrs. Copeland and Worthen, in which no allowance is made for loss of action of the pumps, and the weight of water is the same as used by them, 62.38 pounds; we also assume that the record of coal consumption is strictly accurate, and the fire in the same condition at 12 : 30 M. and 9 : 30 P. M.

Respectfully yours,

WM. E. MORRIS, *C. E.*
SAMUEL MCELROY, *C. E.*

JAS. P. KIRKWOOD, Esq.,
*Chief Engineer, Brooklyn Water Works.*

---

*Report of Messrs. Graff and Smith on the Cambridge Pumping Engine.*

NEW YORK, *June* 29, 1857.

The undersigned having tested the performance of a pumping engine, erected at Cambridge, Mass., for the water works at that place, by Mr. H. R. Worthington, herewith submit the result of that test, with the data from which it was deduced.

As it appeared to be desirable that the conditions of the trial should approach as nearly as possible those of daily working practice, and the boilers not being the subject of exclusive test, it did not seem essential that the time necessary to raise steam in them should be ascertained, nor the fuel for that purpose be noted.

The fires, engine, and boilers, were put into fair working condition, and run until the furnaces required fuel; the amount of the fuel in them was then carefully estimated from measurement, the intensity of the fires closely observed and the steam pressure noted, as also the level of the water in the boilers.

The test was then commenced, and continued for the time mentioned below, care being taken, that the condition of the fires, and the pressure of the steam, should be as near as possible the same at the termination as at the commencement of the trial.

The amount of load upon the pump was ascertained from its area, and carefully observed indications of an Ashcroft pressure gauge placed upon the delivery pipe, an estimated addition being made for the height the water was drawn and forced between the centre of the gauge and the surface of the water in the pump-well, the total pressure thus obtained being thirty-three pounds to the square inch.

The total number of strokes made by the engine amounting to 33,678, were obtained from a counter placed upon it, and confirmed by the average of numerous counts made during the trial.

The length of stroke of the plunger (the average of thirteen careful measurements) was two and one sixth feet.

The total coal consumed was 508.5 pounds: the duration of the test fourteen hours, forty-six minutes.

The data will therefore be as follows:

150.521 inches area × 33 pounds = 4967.193 pounds load upon the plunger.

33678 strokes × $2\frac{1}{6}$ feet = 72969 feet travelled by the plunger.

$$\frac{\text{Then } 4967.193 \times 72969}{508.5 \text{ pounds.}} \times 100 = 71{,}278{,}486 \text{ pounds lifted one foot high}$$

with one hundred pounds of coal.

Previous to the trial the scale upon which the coal was weighed, was placed in a permanent position, examined and sealed: a copy of the certificate of the sealer as to its correctness is appended.

The Ashcroft gauge used, was tested by the manufacturers after the trial, and their certificate of the amount of its error is also attached: the correction of this error has, of course, been made in the data given.

The pump was opened, and the diameters of the plunger and its piston rods measured, during which time the tightness of the delivery valves and plunger under pressure of the column of water was observed; the leakage was very small, that of the valves being just appreciable, of the plunger comparatively very little more.

Effort was made to obtain a measurement of the reservoir, with a view of ascertaining, if possible, the amount of water actually delivered by the pump during stated periods; it was found, however, impracticable to arrive at any satisfactory result, on account of the irregularity of the slope sides, the unknown vacancies behind the stone lining, the absorbent character of the embankment, and the probable leakages of a reservoir of that size and description.

The pumping main is brought up perpendicularly above the surface of the water in the reservoir, which admits of its being seen: observation satisfied us that the water was delivered from it with great regularity, and did not contain any undue amount of air.

The clothings of the steam pipes and cylinders, although good, are not as strictly guarded in that respect as is usual with the Cornish engines in England.

During the experiment the gauges were so placed as to afford an accurate indication of frictional resistance, showing that a head of 6.05 feet was required to overcome the friction of the water through the pumps and also through the main. In other words, the difference between the surveyed height and that indicated by the gauge was 6.05 feet.

The general character of the design and workmanship of the engines is quite creditable, and they appear to be simple, reliable, and durable; during the trial everything worked satisfactorily, and required very little attention from the Engineer.

FREDERICK GRAFF, *C. E.*
ERASTUS W. SMITH, *C. E.*

To MESSRS. J. C. BREVOORT,
JAMES P. KIRKWOOD,
*Committee.*

This is to certify that I have tested the Platform Balance used at the Cambridge Water Works, and find it perfectly correct, according to the state standard.

CAMBRIDGE, *June* 26, 1857. GEO. P. BACKUS, *Sealer.*

BOSTON, *June* 27, 1857.

This may certify that Gauge numbered 8368.55, sold J. J. Walworth & Co., June 26, 1857, was returned to the shop of the American Steam Gauge Co. this day for proof of its correctness, and found to vary one half pound less than the true indication. It is now made to indicate correctly.

H. K. MOORE,
JOHN BUNCE,
*Supts. Am. Steam Gauge Co.*

---

OFFICE OF THE WOODRUFF AND BEACH IRON WORKS,
HARTFORD, *July* 4, 1857.

JAMES P. KIRKWOOD, Esq.,
*Chief Engineer, &c.:*

DEAR SIR: Understanding that you have been further testing the working economy of the Cambridge engine, I beg leave, in behalf of Woodruff & Beach and myself, to request your presence, accompanied by any gentlemen you may think proper to select, to witness a trial of the Hartford engine, on any day, to be named by yourself, after next Wednesday.

The amount of facts recorded at the engine-house induces us to make this request, as they prove that the engine is now producing a regular daily duty greater than has ever been obtained before. The increase is due to lessening the amount of friction of the pump-piston, and by an improvement in the mode of packing, discovered by myself. The change has only been made on two pumps, and the engine now does a duty every day of seventy millions; when the remaining two pumps undergo the same improvement, the duty as a consequence will increase in a like ratio.

The improvements lately made in the working apparatus of this style of engine, together with the improvements in the method of packing the pump-pistons, as alluded to above, give us great confidence in stating that an engine on our plan, now before you, would give a greater result than the Hartford engine, and far exceeding your standard.

Your compliance with the above request, and an early answer, will confer a favor on

Yours, sincerely,

WILLIAM WRIGHT.

It was agreed by Mr. Wright and myself that Messrs. Graff and Worthen should make the second trial of the Hartford engine suggested above. Their report follows.

J. P. KIRKWOOD.

---

*Report of Messrs. Graff and Worthen on the Hartford Pumping Engine.*

NEW YORK, *July* 16, 1857.

JAMES P. KIRKWOOD, Esq.:

According to the request of Mr. Wright and yourself, having made a trial of the pumping engine at Hartford, Ct., we submit the following report of our method of conducting the experiment and the results:

The trial commenced at 10:20 A. M. of the 14th inst., and extended to 4:7 P. M. of the 15th.

Previous to the trial the temperature of the fire-box was found to be 150 deg. Fahr., and there was a vacuum in the boiler. At 9 : 50 A. M. commenced firing with wood, using about one sixth of a cord for that purpose, and then supplying coal to the amount of six hundred pounds, when having raised twenty-five pounds of steam on the boiler, at 10 : 20 A. M., the engine was started and continued to run till the close of the experiment, when the whole of the steam in the boiler was exhausted and the steam gauge showed a zero pressure in the boiler. The experiment was conducted in this manner at the suggestion of Mr. Wright, who wished to show the value of the engine when working intermittently. But in order to afford a comparison with the previous experiments on this engine, and on the engine at Cambridge, Mass., at 2 : 43 P. M. (14th), the state of the fires was observed, and at 2 : 35 A. M. (15th), the fires being then in the same state, the number of strokes of the pump were recorded in the interval, together with the amount of coal used. It will be seen, by comparing the results of these two experiments, that although the duty as shown by the last is larger than for the whole trial, yet it is but a small percentage, and that by running the engine thus to the extreme at the close, the value of the coal used in firing is in a great measure made available. The pressure on the rising main was taken by an Ashcroft gauge, just above the foot-valve ; this gauge compared with a mercury gauge in the engine-room agreed very nearly. The vacuum was also taken by a gauge on the suction-pipe, and the pressure on the pump-pistons is obtained from an average of the observed pressures ; the stroke and size of the pumps were obtained from Messrs. Copeland and Worthen's report:

Average net area of pump piston.......................275.0459 inches.
Length of stroke....................................... 16⅛ inches.

First experiment, from 10 : 20 A. M. (14th) to 4 : 7 A. M. (15th).

Coal consumed:

| | | | |
|---|---|---|---|
| Firing wood, equal to (estimated) | 150 lbs. coal. | | |
| " coal..................... | 600 " " | | |
| Coal, during experiment........ | 2400 " " | | |
| | | ..............3150 lbs. | |
| Residuum................501 lbs. | | | |
| Less ashes...............181 " | | | |
| | | ........................ | 320 |
| Net coal consumed.................................... | | | 2830 lbs. |

Number of strokes of pump........................ 10292
Average pressure.................................. 60.92 lbs.
Duty per 100 lbs. coal...........................65,505,469 lbs. ft.

Second experiment, from 2 : 43 P. M. (14th) to 2 : 35 A. M. (15th), 11h. 52m.

| | |
|---|---|
| Coal consumed | 1800 lbs. |
| Strokes of pump | 6892 |
| Average pressure | 60.93 |
| Duty per 100 lbs. coal | 68,974,549 lbs. ft. |
| Strokes per min., 1st experiment | 9.63 |
| " " 2d " | 9.68 |
| Average pressure on boilers, about | 20 |
| Vacuum | 27½ in. |
| Temperature of hot well | 102 Fahr. |
| " water in rivers | 78 |
| Average lift, about | 120.5 ft. |

The alterations in the machinery since the experiments made by Messrs. Copeland and Worthen, are as follows: the engine has been fitted with a governor; a heater has been introduced into the exhaust, by which the feed-water is supplied hotter to the boiler; the clearances in the cylinders have been reduced, and the pump pistons have been new packed.

It was understood by us that these experiments were intended as a comparison with those made at Cambridge by Messrs. Smith and Graff, and experiment 2d was conducted and the results calculated in the same way as by those gentlemen. It may not be amiss to state that the draft of the pattern from which the cams were made seems to give a somewhat longer stroke than that assumed above, but as this was taken with great care from the machines in the previous trial, and as the actual delivery of water was less than the measured capacity of the pumps, we did not feel at liberty to assume any other without testing by measure and by indicator, as was done before.

Respectfully submitted.

FREDERICK GRAFF,
W. E. WORTHEN.

---

*Report of Messrs. Smith, Graff, and Worthen, on the first Brooklyn Pumping Engine.*

To Messrs. BREVOORT AND KIRKWOOD, *Committee:*

GENTLEMEN: Agreebly to your request, we have made a series of experiments on the pumping engine at Ridgewood, to determine whether it complies with the specifications and contract between the Nassau Water Company and H. S. Welles & Co., contractors; in conformity to the tests for required capacity and duty, agreed upon by James P. Kirkwood, engineer, on the part of the Nassau Water Company, and Wm. E. Morris, engineer, on the part of the contractors, May 30, 1857; and approved by H. S. Welles & Co., contractors, and John H. Prentice, president, and J. Carson Brevoort, secretary of the Nassau Water Company, The method of test above agreed upon is a usual one, and we consider it fair and just to both parties.

The first experiments were made in reference to the "duty" of the engine.

According to the contract and agreement: "The 'duty' of the engines shall not be less than six hundred thousand pounds (avoirdupois) raised one foot high with one pound of mineral coal, and is to be calculated in the usual way, as follows:

"The friction of the water in the pipes shall be ascertained by reliable gauges, and when so ascertained, shall be added to the weight of water due to the height between the pump-well and point of delivery. The weight of the water to be taken at sixty-two and a half pounds per cubic foot.

"The load on the pump piston, thus ascertained, multiplied by the length of stroke, in feet, and by the number of strokes, and divided by the number of pounds of coal consumed during the experiment, equals the 'duty.'

"This test, viz., of duty, to include not less than twenty-four hours' consecutive work of the machine, and the engine while under work shall deliver into the reservoir not less than its proper proportion of twenty millions of gallons in twenty-four hours, to be ascertained by measurement in the reservoir."

The friction of the water in the main was ascertained by an Allen gauge (previously tested), placed at the top of the upper pump, and was found, including the static pressure, to be sixty-one pounds to the square inch, by an average; to which was added 13.077 pounds for the difference of level between the position of the gauge, and the average level of the water in the pump-well during the experiment, or 30.26 feet; making, therefore, a total pressure on the pump-piston of 74.077 pounds per square inch.

No allowance is made for the friction of the water through the pumps and intermediate pipe.

The diameter of the lower pump was found to be a little less than thirty-six inches, while the upper exceeded this measure by about the same amount, the average has been taken at thirty-six inches; the diameter of the lower pump piston-rod was eight and a quarter inches, the area of this section was taken from that of the lower pump, but there was no deduction from that of the upper pump. The sum of the areas of the two pistons is therefore 1,982.3 square inches. And the load on the pistons $74.077 \times 1,982.3 = 146,842.83$ pounds.

The number of strokes during the experiment was 14,923. The average length of stroke, as taken by observation, was 9.83 feet.

In firing, the engine, after having been at work several hours, was stopped, the steam was blown off from the boilers, and the fires were drawn; the firing then commenced anew, an account of the fuel to the end of the experiment was taken; when the engine had exhausted all the steam, and stopped of itself, the grates were raked, and the quantity of coal separate from cinders and ashes was ascertained, and this was credited to the coal used thus:

| | |
|---|---|
| Wood, 1,200 lbs., or equivalent of coal.................. | 600 lbs. |
| Total of coal fired ........................................ | 35,700 " |
| | 36,300 lbs. |
| Less residuum of coal.................................. | 870 " |
| Net quantity of fuel used................................ | 35,430 lbs. |

$\frac{146,842.83 \times 14,923 \times 9.83}{35,430}$=607,982 lbs. ft., or "duty" of 1 lb. coal.

The experiments were commenced January 12th, 1860, at 11 o'clock and 6 minutes A. M., and ended January 13th, 1 o'clock and 11 minutes P. M. Duration of test 26 hours, 5 minutes. The quantity of water delivered at the reservoir was measured both in the reservoir itself, and by a weir as it flowed into the reservoir; the following are the results:

| | |
|---|---|
| By the reservoir measures.......................... | 15,521,719 galls. |
| " weir " .......................... | 15,603,117 " |

The gallon being the New York gallon of 7.8125 to the cubic foot.

By the requirement of the contract, the "duty" was to be 600,000 lbs. feet for 1 lb. of coal, with a capacity of 10,000,000 gallons raised into the reservoir in 24 hours at the same time. It will be seen, therefore, that the machine has complied with the contract requirements of "duty."

It was impossible for us, while at Ridgewood, to get a sealer of weights to examine the platform scales, which we used; but he arrived soon after our departure, and has sent a certificate as follows: "As the index stood on my arrival, 60 lbs. of coal indicated on that index, weighed 59 lbs. avoirdupois." In our calculations no allowance has been made for this discrepancy, as the scales were well balanced during the course of the experiments, but seemed to have been thrown out of adjustment by an upset at the close. If the scales were as indexed by the weigher, throughout the test, about 1.7 per cent. should be added to the "duty" result.

By the contract and agreement, ten "millions of New York gallons of water to be raised into the reservoir, in sixteen continuous hours forms the required capacity of" this machine.

"The water delivered to be measured in the reservoir, making the proper allowance, if any, for absorption, leakage, and evaporation."

The experiments were commenced January 13th, at 7 o'clock P. M., and continued to January 14th, 11 o'clock A. M. Duration of test 16 hours.

The quantity of water was measured as before, both by reservoir and weir, with the following results:

| | |
|---|---|
| By reservoir measures............................. | 10,293,102 galls. |
| " weir " ............................. | 10,095,125 " |

And therefore the machine in *capacity* complies with the contract.

In both series of experiments, allowance on reservoir measures, has been made for absorption, leakage, and evaporation, from observations taken before the commencement and after the conclusion of each test; and during the last experiments, deduction was made for rain-fall from rain-gauge observations.

In regard to the materials and workmanship of the machine, the materials are, as far as we can judge, of excellent quality, and the workmanship superior, and creditable to the builders.

Respectfully submitted.

ERASTUS W. SMITH,
FREDERICK GRAFF,
W. E. WORTHEN.

*January* 17, 1860.

---

## BROOKLYN PUMPING ENGINE.

---

*Mr. McElroy's Notes of the Experiments made to test the Duty, January* 12 *and* 13, 1860.

| | | |
|---|---|---|
| JAN. 12. | Engine running from 5 A. M. | |
| | 10.35′ A. M. | Hauled fires. Counter 560,647. |
| | 10.40′ " | Furnaces wooded with 1,180 lbs. Steam blown off. |
| | 10.45′ " | Started fires. |
| | 11.06′ " | Started engine. Steam, 15 lbs. Well, 8′ 7″. |
| JAN. 13. | 1.09′ P. M. | Engine stopped. Counter 575,612. Well, 7′ 9″. |
| | | Hauled fires and weighed contents. |

Reservoir, Jan. 12. 11.15 A. M. 7′·216.
" " 13. 1.15 P. M. 10′·789.

*Running time*—26 h. 3 min. 1,563 min.

*No. of double strokes*—14,965; average per min. 9.57.

| | | |
|---|---|---|
| *Quantity pumped*—Reservoir prism | 1,943,800 | c. feet. |
| Leakage | 54,262 | " |
| Absorption | 1,938 | " |
| Total | 2,000,000 | c. feet. |
| Rate per 24 lbs. | 14,418,000 | galls. |
| Delivery per stroke | 1,044 | " |
| Capacity " " (for 9′ 10½″) | 1.062 | " |
| Loss of action | 1.69 | per ct. |
| *Equivalent lift* | 170 | ft. |
| *Actual* | 160 | " |

*Fuel account:* Wood, 1,180 = 524 lbs. coal.
Coal charged.. 35,105 "

35,629

" credited.. 856

Total burnt. 34,773 lbs.
Average per hour.. 1,331 "

*Duty:* 2,000,000 c. ft. × 62.5 lbs. = 125,000,000 lbs. × 170 ft. = 21,250,000,000 ft. lbs. ÷ 34,773 lbs. coal = 611,114 ft. lbs.

Friction of engine by cards............8½ ℔ cent.

SAMUEL McELROY,
*Assistant Engineer, B. W. W.*

---

*Notes of Experiment made to test the Pumping Capacity of the Engine.*

| | | | | |
|---|---|---|---|---|
| JAN. 13. | 5.30′ P. M. | Fires lighted. | | |
| | 7.00′ " | Commencement of trial. | Well, 8′9″ | |
| " 14. | 11.00′ " | End of trial. | Well, 7′4″ | |

*Running time*—16 hours ....................................960 min.
*Number of double strokes*—9,708; average per min........10.11
Average speed for four hours, per min..............10.36

*Quantity pumped*—Reservoir prism.............1,291,216 c. ft.
Leakage.............................. 33,392 "
Absorption.............................. 1,192 "

Total......................1,325,800

Per 16 hours ..........................10,357,812 galls.
" 24 " ..........................15,536,718 "

Coal burned..............................26,528 lbs.
" " per hour (18 hours)............ 1,473 "

SAMUEL McELROY,
*Assistant Engineer, B. W. W.*

The Brooklyn engine was built by Woodruff & Beach, of Hartford, Conn., from designs by Mr. Wm. Wright, their Superintendent, and was patented by him, 15th November, 1859.

It delivers at the rate of ten millions of gallons in twenty-four hours, with ease and regularity.

The following are dimensions, &c. :

Diameter of cylinder, 90 inches.
Length of stroke of piston, 10 feet.
Length of beam, 30 feet. Weight of do. 25 tons.
Height of centre of do. above floor, 26 feet 3 inches.
Number of pumps, 2 : stroke of do., 10 ft.
Capacity of do., per double stroke, 137.657 c. ft.
Diameter of working barrel of do., 36 inches.
Diameter of auxiliary barrel of do., 54 inches.
The valves are of the double beat kind.
Double acting air-pumps.
Diameter of do., 3 ft.
Stroke of do., 5 ft.
Diameter of air-chamber, 78½ inches.
Height of air-chamber, 25⅓ ft.
Height of do. above floor, 13 ft. 10 inches.
Return drop flue boilers, three in number.
Diameter of do., 8 ft.
Length of do., 30 ft.
Total weight of engine, boilers, and appurtenances, 440 tons.

Depth of bottom of pump-well below floor, 37 ft. 3 in.
Ordinary depth of water in pump-well when pumping, 7 ft.
Diameter of force tube, 3 ft.
Length of do., 3,450 ft.
Equivalent "lift," 170 ft.

---

*Report of Messrs. Smith, Graff, and Worthen, on the second Brooklyn Pumping Engine at Ridgewood.*

JAS. P. KIRKWOOD, Esq.,
*Chief Engineer, Brooklyn Water Works:*

DEAR SIR : Agreeably to the request in your communication of the 21st instant, and in conformity with the agreement between the parties thereto annexed, we have made an experimental test of the working qualities of the second Ridgewood engine only, the Prospect Hill engine being not yet in condition for trial. The results are as follows :

The test for duty, continued during twenty-four hours, gave for the combustion of one pound of coal an effect of 589,854 lbs. ft., or 10,146 lbs. ft. less than that required by the contract.

The test for capacity, continued during sixteen hours, gave, as measured in the

Reservoir .......................................10,403,414 galls.

By weir.........................................10,275,164 "

thus complying with the contract, whatever measures be adopted.

The experiments were conducted on the 23d and 24th insts., upon the same general principles as on the No. 1 engine ; except that, on account of the fullness of the reservoir, and shortness of time, the experiments on duty and capacity were not entirely distinct, but some eight hours' work has been used in common with both tests ; and the principle of continuous firing has been adopted, instead of drawing the fire as before.

In regard to the character of the workmanship of the machine, and the material of its construction, on the whole, both are extremely satisfactory ; there are some minor details which are deficient, and which should have been attended to previous to the trial, but which may be easily perfected at a very moderate outlay. The engine has been used but nine or ten days, and is subject to all the disadvantages of new work ; we are confident that, after having been used a longer time, and with slight modifications, it will easily come up to the required test of duty, and, as a working machine, prove itself superior in every respect to the first engine.

FREDERICK GRAFF,
W. E. WORTHEN,
ERASTUS W. SMITH.

*October*, 1861.

---

*Notes on the Trial of Ridgewood Engine No.* 2, *October* 23 *to* 25, 1861.

The steam pressure in the boilers ranged from 14½ to 18 lbs. ; the average pressure was about 15½ lbs.

Number of strokes of engine during sixteen hours' trial for capacity, from 2 P. M. October 24, to 6 A. M. October 25, 10,232.

The water was measured in the reservoir, from 9 : 15 P. M. October 23, to 6 A. M. October 25, and at the conclusion of the experiments the loss by leakage was gauged, and allowance made therefor. The quantity thus measured, as delivered into the reservoir during the sixteen hours, from 2 P. M. October 24, to 6 A. M. October 25, was

1,331,637 cubic feet or 10,403,414 galls., N. Y. standard, or, divided by the number of strokes,

130.144 cubic feet per double stroke,

or 1,016.25 galls. "

As a check upon this measure, a weir was made upon the platform over which the water was delivered into the reservoir, and the capacity of the pump was estimated from the heights above the weir, taken every five minutes during four hours, from 2 to 6 P. M., October 24.

Capacity of pumps, as measured by weir:

128.54 cubic feet per double stroke,
or, 1,004.22 galls. "

Formula for weir discharges from Mr. Francis' hydraulic experiments:

$$Q = 3.33\ (l - \tfrac{2}{10} h)\ h^{\frac{3}{2}}$$

$l = 7.25$ ft. h by average 0.993.

Capacity of pump by measure and calculation for a stroke of 9.75 ft.:

134.27 cubic feet per double stroke,
or, 1,048.98 galls. " "

The Ridgewood engine No. 2, not having complied with the contract requirement for duty, as by the report of the engineers, Messrs. Graff, Worthen, and Smith, another trial was deemed necessary, and it was resolved by the Board of Water Commissioners, that their engineer, Mr. Worthen, "be requested to make the second test of said engine, and also to test the engine at Prospect Hill Reservoir."

Time was given for the carrying out of some slight alterations that were found useful by the former trial, and a defect having been found in one of the expansion pipes, it was taken out and a new one put in; and the second trial was commenced on January 9, 1862. The general results of the test were given in a brief note to the chief engineer, Mr. Kirkwood, on the 16th, but the full report (as follows) was not communicated till February 13.

---

*Report of Mr. Worthen on the Second Ridgewood Pumping Engine.*

JAS. P. KIRKWOOD, Esq.:

DEAR SIR: In my previous communication (of the 16th ult.) with regard to the duty and capacity of Ridgewood pumping engine No. 2, as tested by me on the 9th and 10th ult., the results were stated as follows:

606,613 lbs. ft. duty for every pound of coal consumed.

10,554,102 gallons per sixteen hours as the rate of discharge into the reservoir during the twenty-four hours of trial.

These calculations were based on investigations conducted as in the previous experiments on the No. 1 engine, with the exception, that the firing was continuous, and both duty and capacity were tested at the same time.

The load on the pump pistons was ascertained, by observations on the gauge attached permanently to the rising main, and placed beneath the vacuum and steam gauges of the cylinder; to which is added the static pressure due to the height of the gauge above the average level of the water in the pump-well. The boilers were fired up early on the morning of the 9th, and the engine was run intermittently till about 12 M., when everything was ready for trial; the first coal was weighed 12 h. 30 m., and record was then taken of every shovelful of coal that was thrown on the grates, and upon what grates it was thrown. The coal was weighed in lots of three hundred pounds each; record was taken of the time when each lot was cleaned from the floor, and from the number of shovelfuls in each lot, the weight of each shovelful is estimated and plotted on the profile hereto annexed.* The horizontal lines represent hours, a division of ten to the hour, or six minutes to each small division; the perpendicular divisions represent the weight of coal, one hundred pounds to each small division. By observation of a profile thus constructed, it will be seen how much coal is consumed from hour to hour, and with what care the firing was conducted. During the whole experiment the fire, water, and steam, were kept as nearly constant as possible.

The firing was continued and coaling noted till after 4 o'clock P. M. on the 10th ult., but the time selected for the experiment has been the twenty-four hours, from 2 P. M. 9th, to 2 P. M. 10th; the quantity consumed has been taken from the profile. As in the experiments on No. 1, the quantity of coal in the ashes and cinder, during the whole firing, has been taken, and the percentage due to the quantity consumed during the test has been deducted. This, with a small deduction of coal, for an excess of half an inch of water in the boilers, at 2 P. M. 10th, over that at 2 P. M. on the 9th, has been taken as the net consumption on which the above calculation of duty has been based.

As in these last tests the responsibility for the accuracy of the experiments rested upon myself alone, and that the data obtained might be open to inspection of parties not present, in addition to observations of the gauge, cards were taken from indicators at the top of both pumps, in the positions *I*, *I'*, shown on Plate I., two at each place every hour; a few of the cards are annexed, but the average was taken for the whole twenty-four hours, fifty from each pump.

From the average thus taken the duty has been calculated, as will be explained hereafter; and although it is somewhat above the duty as established by the gauge, and heretofore given, I think it is the more reliable and correct, as the data are of record and not of observation, and contain the friction, not previously noted by the gauge, of the main beneath it.

The coal used was Buck Mountain, and the residue of fine coal which fell through the grates unconsumed, and was deducted in the previous calculation, has not been deducted in this, as this residue, although it might be useful for some purposes, will never be of enough account to screen for firing under these boilers.

Indicator cards were taken at each end of the steam cylinder every hour during

* The profile accompanying the report has not been printed, but its construction will be understood from the partial one now attached to the report of the Prospect Hill engine.

the whole trial, and an average of them is given for the twenty-four hours of the trial for the purpose of comparing the steam load with that of the pumps, and the power absorbed in the working of the feed and air pumps, the valve geer, and friction of the machinery.

The dimensions of the pumps were taken in the previous trial, and were found to be thirty-six inches in diameter.

| | | |
|---|---|---|
| The area is, therefore | 1,017.88 | sq. in. |
| The pump piston-rod, 8¼ diameter, area | 53.46 | " |
| Net area of piston | 964.42 | " |

From observations on the 9th and 10th insts., during the twenty-four hours:

| | | |
|---|---|---|
| Average pressure at lower pump indicator | 76.07 | lbs. |
| " vacuum " " " | 6.48 | " |
| " pressure at upper pump indicator, up-stroke | 65.04 | " |
| " " " " " down-stroke | 67.01 | " |
| Counter at 2 P. M. 9th inst. | 96,697 | |
| " 2 " 10th inst. | 112,183 | |
| Number of double strokes during the twenty-four hours | 15,486 | |
| Average length of stroke | 9.74 | feet. |
| Temperature of water in pump-well | 42° | Fah. |
| " " hot well | 70° | " |
| Average pressure in boilers | 18 | lbs. |
| " temperature in boilers by Pambour's tables | 257° | Fah. |
| Water in boilers at 2 P. M., 10th, ½ inch higher than at 2 P. M., 9th inst., or by calculation water | 1,869 | lbs. |
| Coals used to 2 P. M., 9th inst., by profile | 2,420 | " |
| " " 2 " 10th inst., " | 41,530 | " |
| Coal used during the twenty-four hours of test | 39,110 | " |
| " required to raise 1,869 lbs. of surplus water in the boilers from 70° to 257°, $\frac{1869}{53.4}$ = | 35 | " |
| Net coal used during the twenty-four hours' test | 39,075 | " |

WATER LOAD IN THE UP-STROKE OF THE STEAM PISTON.

The water above the lower pump piston exerts a pressure on the piston equal to the sum of the pressure, as shown by the lower indicator, and of that due to the height of the indicator above the piston at its half-stroke.

Beneath the piston we consider the vacuum to be that due to the height of the piston at half-stroke above the average level of the water during the experiment. This is not strictly correct, as will be easily understood by looking at Plate I.;

nor is the pressure above the piston at half-stroke exactly that due to the average load, in consequence of the height of the water in the well; the latter is a little in excess, while the allowance for the vacuum is a little too small; the errors very nearly balance, as we find by calculation.

As the lower pump piston is raised, the upper pump piston descends, and the upper piston rod acts as a pole plunger, and displaces water equal to its section multiplied by its stroke, under an average pressure equal to that due to the position of the piston at half-stroke. This pressure is determined from the upper indicator, and the height of the indicator above the piston at its half-stroke. Hence we estimate the pressure on the pump pistons in the up-stroke of the engine, as follows:

| | | |
|---|---|---|
| Average pressure at lower indicator | | 76.07 lbs. |
| Height of indicator above piston at half-stroke—Plate I.—10 ft. pressure, $\frac{10}{2.31}$ = | | 4.33 " |
| | | 80.40 " |
| $80.4 \times 964.42$, net area of piston pressure on top of lower pump piston | | 77,539.37 " |
| Height of lower pump piston at half-stroke above average level of water during experiment | 2.12 ft. | |
| Vacuum due to this height, $\frac{2.12}{2.31}$ = | 0.92 lbs. | |
| $0.92 \times 1017.88$, area of bottom of pump piston | | 936.45 " |
| Total load on lower pump piston | | 78,475.82 " |
| Average pressure of upper indicator at down-stroke of upper pump | 67.01 lbs. | |
| Height of indicator above piston at half-stroke 16.08 ft. | | |
| Pressure due at this height, $\frac{16.08}{2.31}$ | 6.96 " | |
| Total pressure on piston at half stroke | 73.97 lbs. | |
| $73.97 \times 53.46$, area of piston rod | | 3,954.43 lbs. |
| Total water load on up-stroke of piston | | 82,430.25 " |

#### WATER LOAD IN THE DOWN-STROKE OF THE STEAM PISTON.

The upper pump piston is raised, and the average pressure exerted on the upper surface of the piston is estimated as equal to the sum of the pressure, as shown on the upper indicator, and that due to the height of the indicator above the piston at half-stroke. Beneath the piston the vacuum is estimated by the average vacuum, as shown on lower pump indicator, and the height of the piston at half-stroke above this indicator. The lower pump piston rod in the down-stroke of the engine, is partly drawn down by the vacuum, and partly forced beneath the level of the water in the pump well;

calculation has been made of the relief due to the excess in the vacuum, but is now neglected as of too little amount.

| | | |
|---|---|---|
| Average pressure at upper indicator in up-stroke of upper pump........................................ | | 65.04 lbs. |
| Height of indicator above piston at half-stroke. | 16.08 lbs. | |
| Pressure due to this height, $\frac{16.08}{2.31}$ ........................ | | 6.96 " |
| | | 72.00 " |
| $72 \times 964.42$ net area of piston........................... | | 69,438.24 " |
| Average vacuum by lower indicator ............... | 6.48 lbs. | |
| Height of upper pump piston at half-stroke above indicator ........................... 4.32 ft. | | |
| Vacuum due to this height, $\frac{4.32}{2.31}$................ | 1.87 " | |
| | 8.35 | |
| $8.35 \times 1,017.88$, area of bottom of piston.................... | | 8,499.30 lbs. |
| Total load on upper pump piston........................ | | 77,937.54 " |
| " " lower pump piston ........................ | | 78,475.82 " |
| " " upper pump piston-rod..................... | | 3,954.43 " |
| Total water load............................ | | 160,367.79 " |

The total water load, multiplied by the length of stroke and by the number of strokes, and the product divided by the coal consumed during the strokes, will be the duty of the engine; thus:

$$\frac{160367.79 \times 9.74 \times 15486}{39075} = 619,037 \text{ lbs. ft.}$$

for each pound of coal consumed during the experiment.

Capacity from Messrs. Graff, Smith, and Worthen's report on experiments made 23d and 24th October, 1861: 10,403,414 gallons discharged into the reservoir in sixteen consecutive hours.

In my previous communication, the test of capacity was estimated during the whole duration of the duty test, and referred to the required time of sixteen hours; but, during the first sixteen hours, the average number of strokes and discharge of pump was somewhat larger than the subsequent eight hours.

During the first sixteen hours of the experiment there were pumped into the reservoir, as there measured, 10,652,366 gallons.

For a number of hours consecutively, the average number of strokes per minute was very nearly eleven, or at the rate of about 10,800,000 gallons per sixteen hours.

Capacity of pumps per double stroke:

By calculation, 1,047.5 gallons

Reservoir measure, 1,022.3 gallons.

By the average of the indicator cards, taken from both ends of the steam cylinder every hour during the twenty-four hours of the test, the pressure was found to be 15.475 lbs.

The diameter of the cylinder, as taken from Messrs. Woodruff & Beach's plans, was 85 in., or 5,674.51 sq. in, area; 15.475×5,674.51=87,813 lbs. average load on steam piston during each stroke.

The average load on the pump pistons is:

$\frac{160368}{2}$, or 80,184 lbs.,

or a little less than nine per cent. less load on the pumps than on the steam piston. No. 2 engine is therefore fully equal to No. 1 in respect to the economy of working and friction of parts; as, in addition to the work done by No. 1, the feed-pump to the boilers is connected with No. 2, while the feed to the No. 1 boilers is independent of the engine, a Guild & Garrison pump being made use of for this purpose.

By calculation by volumes from the steam cards, the evaporation of the boilers was found to be about 8¼ lbs. of water per pound of coal; a result very nearly equal to that found by Mr. Copeland and myself in our experiments on the Hartford pumping engine.

The usual guards, for accuracy in the experiments, have been made, by testing and comparing the indicators and scales; and they were found to be in such condition as to require no correction in the data obtained.

The results deduced, I trust, will be found consistent with the data, and the method of the investigation satisfactory to yourself.

Yours truly,

W. E. Worthen.

*February* 13, 1862.

DESCRIPTION OF RIDGEWOOD PUMPING ENGINES.

The Ridgewood Pumping Engines, Nos. 1 and 2, in plan and general construction, are almost identical, but differ somewhat in the details of the valve gear, and in the dimension of the steam cylinder.

The figure below represents, in outline, a sketch of No. 2 Pumping Engine, consisting of a common double-acting and condensing steam engine *S*, whose piston rod is attached to one extremity of a balanced beam *B*, working on a main centre *m*; to the opposite extremity of the beam is attached the pump rod of the upper pump *p′*, while the piston rod of the steam engine, extending through the bottom of the steam cylinder, forms the pump rod of the lower pump *p*. Both pumps are single-acting bucket pumps ; the upper one

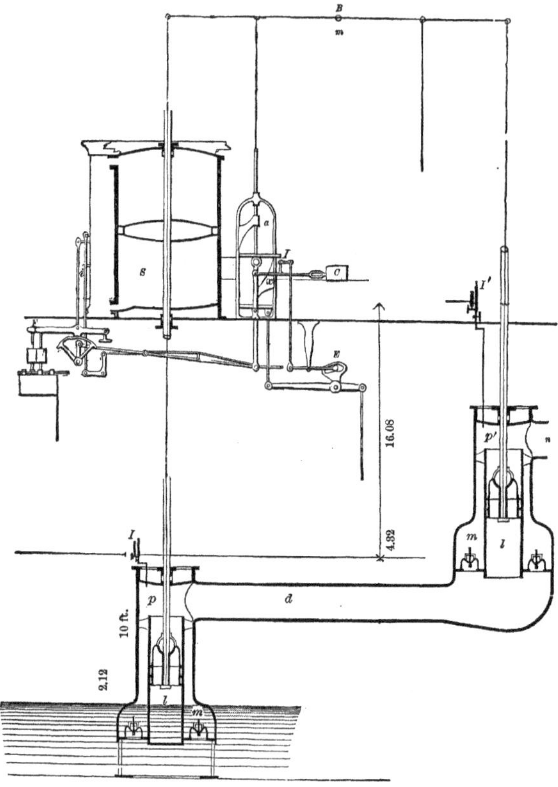


drawing its water directly through the bucket of the lower pump, and the valves in the annular space around it ; and the lower pump discharging through the bucket of the upper pump and its annular space.

The valve motion is worked by means of the inclines in the frame *a*, suspended from the beam *B*, and the water cylinder *C*. The general position and connection of parts can only be seen in the figure, but all the details are illustrated on a larger scale in Plate No. 35, and will be fully explained hereafter. *I*, *I′*, represent the position of the indicators during the test experiment, and the figures are their distances, in feet and decimals, from the valve lines of the upper and lower pump-buckets, when at the centre of the stroke, and of the latter above the average level of the water in the well during the experiment.

The air-pump is worked from a connection with the beam *B*, and with the condenser and pumps, is placed between two strong longitudinal walls, extending nearly to the floor, *F*, of the engine room. The cast iron frame supporting the main pillar block, rests on the top of these walls, and the steam cylinder, air-pump, condenser, and main pump, are secured to strong cast iron cross girders, giving as much access as possible for the inspection and repair of parts. The hand-wheels, for the control of the engine, are at the side of the cylinder, on the level of the floor, *F*.

## VALVE GEAR.

In Plate No. 35, is represented on a large scale the gearing for operating the steam and exhaust valves of the engine—the steam piston is shown at half-stroke on its downward movement, the various parts of the gear being drawn in exact accordance with that position and direction, the upper steam and lower exhaust valves being open. On Plate No. 36, is represented on a still larger scale a horizontal and vertical section of the water cylinder *C*.

As already explained, the frame or yolk *a*, with the inclines *b* and *b′* on its sides, receives its motion from the beam and moves in the same direction as the steam piston, and is for the purpose of closing the steam and exhaust valves; the levers *c* (two) with the rolls on the upper ends transmit the motion from the inclines through the rod *d*, arm *e* and *f*, rod *g*, and arm *h*, to the exhaust valve rock shaft—the exhaust valve being closed when the lever *c* has made half its movement, or reached its vertical position. The steam valves are closed by the same frame or yolk *a* through the rod *i*, lever *j*, arm *k*, rods *l*, *l*, and rock arm *m*, communicating motion to the closing arm *A* of steam valve, The closing arm *A* is in the form of a segment—the face being made of two curves, the difference in their radii being just equal to the lift of the steam valve, this segment is adjustable by hand—that is, the closing face can be brought sooner or later under the toe of the lower lever *B* or upper steam valve, performing the closing or "cut off" at the point required.

The "water" cylinder C, is used exclusively for *opening* the steam and exhaust valves, and is furnished with a piston, admission and exhaust ports, valves, &c., like any steam cylinder. The piston rod is attached to a cross head and connected to the levers *C* by the rod *D*. The admission and exhaust valve is operated from the double curved arm *E*, on the rockshaft *E*. The slot in the arm *E* is composed of two curves joined together in the vertical centre by an inclined slot; the difference in the radii of the curves being equal to the movement of the valve, and the length of the incline determines the time of action, on the end of the lever *G*, is a roll revolving freely on a journal, and fitting the slot in the arm *E*, the vibrating motion of the arm raising or depressing the roll from one curve to the other, imparting a like motion to the other end of the lever *G*, and through the rod *H*, and right-angled arm *I*, to the valve rod *K*. It will be observed that this valve motion is intermittent.

There is another valve in the "water cylinder chest" which we call the supply valve—as it opens and closes the communication between the water chest and the rising main, and is operated from the roll levers *c*, through the connecting rod *L*, and lever *M*, attached to the valve rod ; this motion to the supply valve is also intermittent, and made by the slot in the rod *L* and lever *M*. The ends of the lower levers *B* and *B′*, of the steam valves are connected to weighted plungers *N* and *N′* working in small open cylinders, one for each steam valve, the gravitation of the weighed plungers open the steam valves through the rods *O*, and lever *P*, attached to the valve-stems, and the time taken to open the valves wide—being regulated by the velocity of the water forced out of the cylinder, a small cock being fitted to each cylinder, for this special adjustment.

This completes the description of the various parts of the gearing and their separate duties ; what follows will be an effort to show the combined action of the whole in relation to the motion of the steam piston, and the successful performance of the labor required of it, but from the nature and novelty of the gearing it is scarcely possible to describe it by a diagram so as to be clearly understood by all that may chance to read it.

The steam piston having reached midway on its downward stroke, of course the frame or yolk (*a*) is also at mid-position in the same direction, and the upper steam valve open—its weighted plunger unlatched and at the bottom of its cylinder ; the lower exhaust valve is also open leaving free communication between the under side of the steam piston and the condenser. These are the positions of the parts as illustrated on Plate 35. Now, as the "cut off" usually takes place at or about half stroke in this engine—it will be seen by reference to the sketch that the closing part on the periphery of the arm *A* is just entering under the toe *Q* of the lower lever of the upper steam valve—and a very little more movement of the arm *A* will close the valve, that is, the lower lever will be lifted by the action of the arm, transferring its motion through the rod *O*, and lever *P*, to the valve stem, forcing it downward and closing the valve, at the same time the lower lever *B*, on its lift carried with it, its weighted plunger to the top of its working cylinder *R*. The latch bolt *s*—shown on the sketch as withdrawn—enters a socket in the plunger by the action of the spring on its back, and holds the plunger there until it is time to open the upper steam valve again on its next downward stroke—the closing arm *A* moves on without producing any further motion to the lower lever *B*, its toe simply resting on the curve of the arm as it passes under it.

The steam valve being now closed, the balance of the down stroke is made by the expanded steam—the frame or yolk *a* is descending, and the water cylinder piston *C* is at full stroke toward steam cylinder, the admission valve *K* of the former has just accomplished its half movement—that is, it is square over its cylinder ports, as will readily be understood by a glance at the position of the various parts employed in producing its motion ; when the frame or yolk *a* has descended so that the lower end of the upper incline *b*, comes in contact with the roll *t*, on the lever *c*, the water-cylinder admission valve *K* has completed its movement, opening the back port to the exhaust passage, and removing the water on the back of the piston, and leaving the front port open ready for the admission of pressure on the front side of the piston. The motion of this admission valve ceases now, or at least until the frame *a* has arrived at the same position on the up-stroke, when a similar but of course a reverse motion takes place. This cessation of motion to the admission valve is accomplished by the rock arm *E*, having moved its curved slot over the roll, the movement to the valve attachments being produced only (as before described) while the inclined part of slot in the arm *E* is passing under the roll of the lever *G*.

The frame or yolk *a* still descending under the operation of the expanding steam, the upper incline *b* has forced the lever *c* over to its vertical position or half movement, closing gradually the lower exhaust valve through the various connections, rock arms and toes. The duty of the upper incline *b* ends here for the down-stroke ; at this same instant the admission valve covering the supply port between the water-cylinder chest and rising main is opened by the action of the rod *L*, and lever *M*, and water under pressure permitted to enter the already open front port of the water cylinder, carrying the water piston to the end of its stroke, and with it the lever *c*, through the connecting rod *D*, which movement opens the upper exhaust valve, and permits the steam that has just expended its power on the down-stroke to escape to the condenser, by this same movement the short arm *V* comes in contact with the end of the slot in the latch bolt of the lower steam valve, withdrawing it, and allowing the lower steam valve to be opened by the gravitation of the weighted plunger *N*, through similar rods, levers, &c., as described for the upper steam valve. The engine is now reversed and commencing its upward stroke all the movements just enumerated for the down-stroke, will be again repeated on the up-stroke, but the direction reversed, the lower incline *b*, on the frame *a* performing a similar duty on the upper stroke as the upper one did on the down-stroke, and so on continually.

PUMPS (see page 130).

The lower pump *p*, is placed in a well below the engine-room floor, and directly under the steam cylinder ; the upper pump, *p′*, is also beneath the floor, and connected directly to the opposite end of beam. This pump has a nozzle, *n*, to which the forcing main is joined. Each pump is constructed of two barrels, *l* and *m* ; the inside one, *l*, being the working barrel, fitted with a bucket and double-beat valve, and the annular space between them being connected to the suction by eight double-beat valves.

To the novel form of these pumps is due in a great measure the efficiency of the machine ; for by this combination of the two barrels there is presented to the moving column of water through the valves a net area much larger than that of the working barrel itself, thus allowing the velocity of the column to conform to that of the moving bucket, and in this way reducing the friction to a minimum.

Their operation is as simple as it is effective, and is described in a very few words as follows:

The lower pump on its up-stroke lifts the water above its bucket through the connecting pipe, *d*, and through the annular valves and valve of descending bucket of upper pump, at the time charging itself below its bucket by suction. On its down-stroke, the upper pump is lifting the charge above its bucket, and filling the space below it by suction; the lower pump again ascends, performing the above operation, each pump moving on its up-stroke a column of water about 36 inches in diameter, ten feet toward the reservoir. The action of the pumps is well illustrated by the cards from the Indicators. (Page 135.)

### BOILERS.

The boilers are three in number and of the return drop flue variety:

Length of boiler.............................. 30 feet.
Diameter.................................... 8 "

Two fire-boxes to each boiler:

Length of grate......................... 6 ft. 8 inches.
Width of each grate..................... 3 ft.

Four upper flues to each fire-box:

Length of upper flues..................... 21 ft. 6 inches.
Diameter................................ 19 "

Sixteen lower flues to each boiler:

Length of lower flues..................... 18 ft. 6 inches.
Diameter................................ 8½ "

### EXPLANATION OF PUMP INDICATOR CARDS.

The indicator *I* on the lower pump gives a complete card, but in determining the load on the lower pump piston, the pressure above the piston is taken from the upper line of card (76.07 lbs.), to which is added the pressure due to the height of the indicator (ten feet) above the piston, at half-stroke, and the vacuum below the piston is taken from the height (2.12 feet) of the piston at half-stroke above the average level of the water in the well during the experiment.

INDICATOR CARD FROM STEAM CYLINDER.

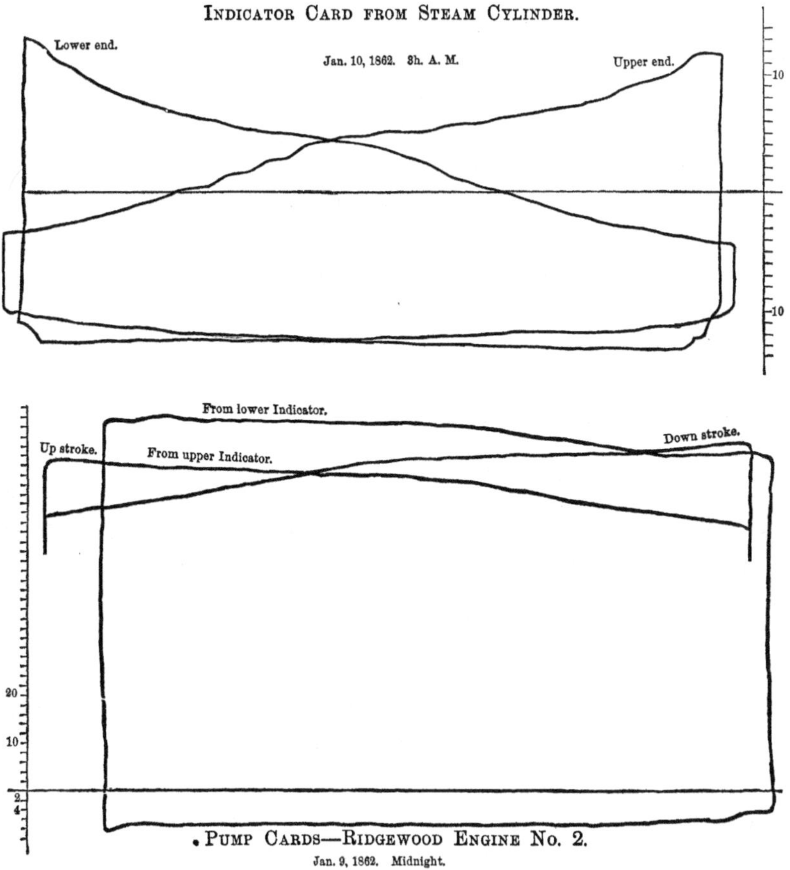


PUMP CARDS—RIDGEWOOD ENGINE NO. 2.
Jan. 9, 1862. Midnight.

The indicator *I′* from the upper pump gives a card in which two lines of pressure are represented; that from the right to left is due to the up-stroke of the lower pump; that from the left to right is due to the up-stroke of the upper pump, and is used (65.04 lbs.) in making up the load on the upper pump piston, to which is added the pressure due to height (16.08 feet) of the indicator above the piston at half-stroke, the vacuum below the piston is taken from the vacuum (6.48 lbs.) as shown on the card from the lower indicator, to which is added the vacuum due to the height (4.32 feet) of the upper pump piston at half-stroke above the indicator *I*. (See page 130.)

*Report of* MR. WORTHEN *on the Pumping Engine at Prospect Hill.*

JAMES P. KIRKWOOD, Esq., *Chief Engineer, Brooklyn Water Works.*

DEAR SIR: Having made a series of experiments, to test the capacity and duty of the pumping engine at Prospect Hill, I respectfully submit the following report of the manner of conducting the experiments, the method of calculating, and the results obtained.

The pumps were ready for operation, and were started, for the first time, on the 5th ult. So long time had already been consumed in the construction of the machine, and so important was it to have the trial and report as soon as possible, that it was decided to go on with the test without any preliminary running of the engine. A week was consumed in putting on the small pump which returns the injection water to the rising main, and arranging some small matters for the placing of the indicators, and on Tuesday, the 13th ult., the trial was commenced. Up to this date, the pumps had not been run, in the aggregate, twenty hours, and no data had, or could have been obtained, of the best circumstances, as to pressure of steam and speed, from which to obtain the required duty; I therefore decided to run the engine during the balance of the week, varying the pressure of steam, the velocity of the engine, and the pressure upon the induction side of the pumps. The results have been tabulated, and may be of use for future reference.

The experiments were continued until 9 A. M. of the 17th ult., when, having filled the Reservoir to the height of nearly one foot above that ever contemplated, and finding that the engine had already complied with the requirements of the contract, the fires were drawn.

The methods of firing and of noting the coal consumed were similar to that adopted in the experiments on Ridgewood Engine No. 2. The boilers were fired up early on the morning of the 13th ult., but the experiments were not commenced till noon: the counter of the engine was then taken, the height of the water in the Reservoir and the quantity of coal, as soon as it was required. The coal was weighed in lots of one hundred pounds each; record was taken of the time of each firing, and the quantity used; a profile was then made, on cross-section sheets, the horizontal lines representing the times, and the perpendicular lines the quantity of coal. And from the profile can be readily seen the quantity of coal consumed from hour to hour.

As in all previous experiments, the quantity of unconsumed coal in the ashes and clinkers has been deducted from the quantity of coal fired; the same allowance has been made in these experiments, but the deduction is so small, that were it omitted, the engine would still far exceed its requirements.

The quantity of water discharged into the Reservoir is deduced from heights taken in the glass tube in the Gate-House, and from tables of capacity furnished by Mr. Lane.

The work done at the pumps is calculated from the average pressures from indicator cards taken every half hour; from the number of revolutions, as shown by the engine counter; and from the measured length of stroke and diameter of the pumps; to which is added the pounds feet of work done at the small plunger pump, discharging the injection water into the rising main. No account is taken of the work done in supplying the boiler feed.

The accompanying figure illustrates the points from which pressure cards were taken. The indicator, by means of cocks, could be connected with either of the points *a*, *b*, or *c*. When in connection with *a*, the connections with the other points being shut off, the line shown on the indicator was the pressure with which the water was forced into the first pump *P*; when in connection with *b* only, two lines are shown on the card, the upper being the pressure of the water as it is lifted by the bucket of the first pump, the lower line the pressure of the water as it is forced into the second pump *P′*, while it makes its up-stroke. When in connection with *c* only, the indicator gives nearly a single line upon its card, the pressure on the rising main at this point.

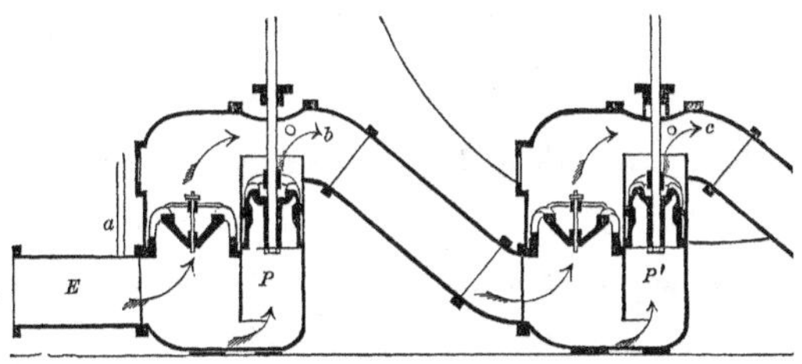


The load at the first pump is estimated by ordinates on the cards between the lines as given from *a*, for the lower line, and the upper line as taken at *b*; while the load of the second pump is estimated from ordinates between the lower line taken from *b*, and the line from *c*, for the upper line. The lines are marked in light and heavy line on the accompanying cards. From these cards thus taken every half hour, and pressures averaged, a profile has also been plotted, and from this profile the average pressure on the pumps from hour to hour has been estimated, measured, and tabulated, to supply the data for calculation.

The dimensions of the pumps were found to be as follows:

| | | |
|---|---|---|
| Diameter of lower pump | 20.3125 | inches. |
| " " upper pump | 20.1875 | " |
| Or an average of | 20.25 | " |
| Stroke of lower pump | 3.4685 | feet. |
| " " upper pump | 3.4633 | " |
| Average taken at | 3.466 | " |

Pump returning injection to the main:

| | | |
|---|---|---|
| Diameter of plunger | 8.0 | inches. |
| Stroke | 1.604 | feet. |

Diameter of piston rod of large pumps, three inches, which is to be taken from the area of the lower pump, but acts as a plunger pole in the down-stroke of the upper pump.

| | |
|---|---|
| Area of both pumps | 644.126 sq. in. |
| Less piston rod | 7.069 " |
| Net area | 637.057 " |
| Area of return pump | 50.265 " |

It was my intention to have selected as the test of duty, the twenty-four consecutive hours which gave the highest result; but, after making my calculations, I thought it would be most satisfactory if the result of the whole trial were given, and the maximum of ten hours' trial, near the last of the experiments.

| | |
|---|---|
| Coal fired during the experiments, from May 13th, at noon, to May 17th, at 9 A. M., 93 consecutive hours | 14,180 lbs. |
| Add for difference of level of the water in the boiler, at beginning and at end of trial | 40 " |
| | 14,220 " |
| Less coal in ashes | 383 " |
| Coal consumed | 13,837 " |
| Average pressure from pump cards | 29.96 " |
| " " " plunger pump | 40 " |
| No. of strokes | 129,557 |

$$637.057 \times 3.466 \times 29.96 \times 129{,}557 \div 13{,}837 = 619{,}381$$
$$50.265 \times 1.604 \times 40 \times 129{,}557 \div 13{,}837 = 30{,}196$$

Duty of 1 lb. of coal............649,577 lbs. ft.

The requirement of the contract was 600,000 lbs. ft. The engine therefore exceeds her requirements, on a test, never contemplated in the contract, of 93 hours instead of 24 hours.

| | |
|---|---|
| The duty estimated in a similar manner for 10 hours, from 7 P. M., May 16th, to 5 A. M., May 17th, was found to be | 684,042 lbs. ft. |
| The capacity, as gauged in the reservoir, for the time required by contract, 16 hours, from 12 P. M., 13th, to 4 P. M., 14th ult., was found to be | 3,068,304 gallons. |
| Requirement of contract | 2,500,000 " |
| The capacity of pumps, as gauged at the reservoir, for 78 hours, from noon, May 13th, to 6 P. M., May 16th, was found to be | 112,036 " |

no allowance being made for leakage.

The number of strokes during the 93 hours, or whole trial, was .................................... 129,557

129,557 × 112,036 = 14,557,027 gallons for 93 hours, or at the rate of

For 1 hour ................................ 156,557 gallons.

For 16 hours ............................. 2,504,435 "

The results above deduced were entirely unexpected, but would be easily exceeded by the same pumps, were it attempted to obtain the highest duty and capacity during sixteen hours only, by continuing at a uniform rate of speed and pressure. The position of so large pumps, upon a rising main, drawing their supply under a pressure of from 7 to 18 lbs. per square inch (varying with the draft in other parts of the city), is unusual; but, from the indicator cards, and general working of the engine, it appears that the pumps are as steady in their action as if they drew from a well, and much more economical in the consumption of coal, as the load is reduced by the amount of pressure in the main.

In some of the details there are a few trifling errors, which for the most part can be readily obviated in the after working. As a whole, I consider the machine satisfactory in its construction, results, and location.

Yours, truly,

*June* 3, 1862. W. E. WORTHEN.

## DESCRIPTION OF PROSPECT HILL PUMPING ENGINE.

PLATE II. represents a section in outline of the pumps, and an elevation of the pumping engine at Prospect Hill.

PLATE II.

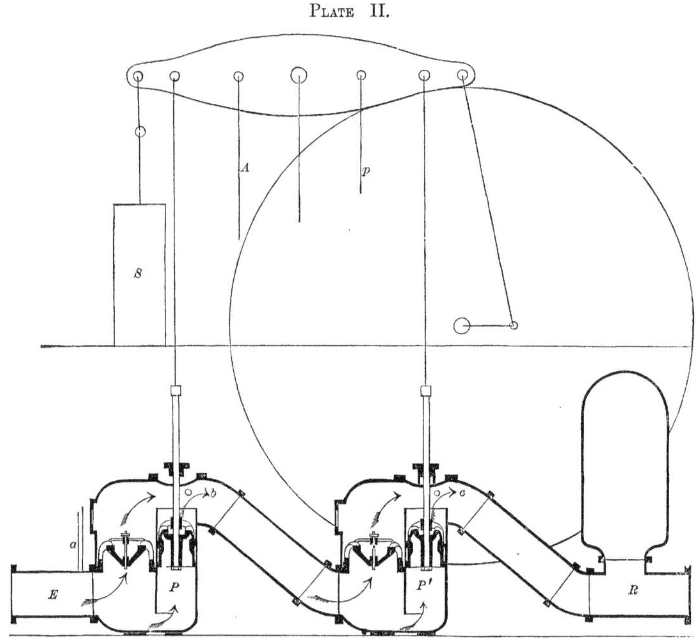


It will be observed that the engine is of the crank and fly-wheel variety, and in this respect is entirely different from the engine at Ridgewood. The steam cylinder *S* is fitted with slide valves, and a cut-off controlled by a governor constructed under Wright's patent. The speed during the experiments was varied by weights suspended to the governor rods. The pumps are constructed on the same general principles and mode of action as the pumps at Ridgewood. They are two pumps attached to opposite sides of the working beam; they have valves in their buckets, and in channels at the sides of the pump. The pumps are placed in a branch-main, and the water flows into and through the air pumps under a considerable head, variable with the draft upon the mains in other parts of the city. There are two air-chambers: one shown at *R* on the rising main, the other, and a somewhat larger one, is connected by a branch pipe with the induction pipe at *E*.

At *A* are the connections with the air pump. At *p* the connections of a small single acting plunger pump to supply the boiler feed, and return the injection water to the main.

By *P* is denoted the lower pump; by *P'* the upper; *a*, *b* and *c*, represent the apertures in connection with the indicator as explained in the report.

## DIMENSIONS.

### *Steam Cylinder.*

| | |
|---|---|
| Length of stroke | 4 ft. 6 in. |
| Diameter of cylinder | 24 inches. |
| Diameter of piston rod | 3¼ " |

### *Pumps.*

| | |
|---|---|
| Length of stroke (average) | 3.466 ft. |
| Diameter of barrels | 20¼ inches |
| Diameter of piston rods | 3 " |

### *Pump to turn injection Water into the Main.*

| | |
|---|---|
| Length of stroke | 1.604 feet. |
| Diameter of plunger | 8 inches. |

### *Fly Wheel.*

| | |
|---|---|
| Diameter | 20 feet. |
| Length of crank | 27 inches. |

### *Boiler—One, Drop Flue.*

| | |
|---|---|
| Length of shell | 18 feet. |
| Diameter of shell | 6 " |
| Length of fire grate | 5 " |
| Width of each fire grate during trial | 2 ft. 2½ in. |
| Number of upper flues | 4 |
| Diameter of upper flues | 13 inches. |
| Length of upper flues | 11 feet. |
| Number of lower flues | 9 |
| Diameter of lower flues | 7 of 9 in., 2 of 7 in. |
| Length of lower flues | 9 ft. 3 in. |

## EXPLANATION OF COALING PROFILE.

The profile of the coaling for 17 hours, will serve as an explanation of the form in which the register of coal consumed has been kept and plotted during the late experiments at Ridgewood and Prospect Hill. The hours selected have been taken rather than those at the commencement of the experiments, as profiles of steam and water pressures are given for the same period.

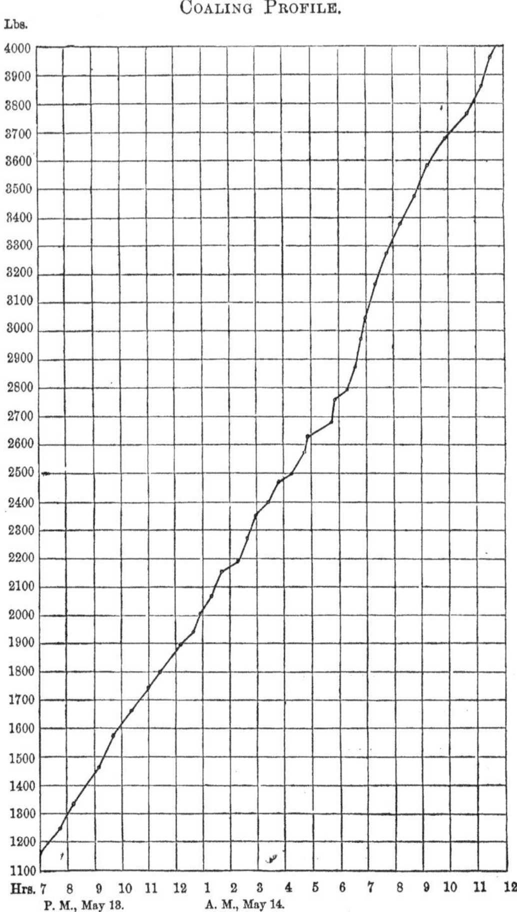


Coaling Profile.

The firing was commenced on the morning of May 13, and the fires and water were got into the state in which it was determined to keep them as nearly as possible uniform. The coal was first noted at 0*h.* 14*m.* P. M., when 100 lbs. were thrown on one of the grates; at 1*h.* 2*m.*, 100 lbs. on the other; at 1*h.* 55*m.*, 60 lbs. on the first grate, and at 2*h.* 32*m.*, 20 lbs. on the second. In this way the quantity of coal was taken every time any was put on, and on which grate thrown. The fire box was divided in two by a brick wall, to maintain a more even fire. The coal was weighed in lots of 100 lbs. each, and the amount at each firing was then estimated. At

| | | | |
|---|---|---|---|
| 6.56 | P. M. | The total quantity fired was | 1,155 lbs. |
| 7.43 | " | 90 lbs. | 1,245 " |
| 8.12 | " | 100 " | 1,345 " |
| 9.03 | " | Cleaned fire No. 2. | |
| 9.10 | " | 120 lbs. | 1,465 " |
| 9.47 | " | 115 " | 1,580 " |
| 10.27 | " | 80 " | 1,660 " |
| 10.58 | " | 80 " | 1,740 " |
| 11.30 | " | 60 " | 1,800 " |
| 12.00 | " | Cleaned fire No. 1. | |

| | | |
|---|---|---|
| 12.06 P. M. | 100 lbs. | 1,900 lbs. |
| 12.37 " | 50 " | 1.950 " |
| 12.56 " | 50 " | 2,000 " |
| 1.25 A. M. | 70 " | 2,070 " |
| 1.47 " | 80 " | 2,150 " |
| 2.22 " | 50 " | 2,200 " |
| 2.42 " | 70 " | 2,270 " |
| 3.02 " | 90 " | 2,360 " |

and so on; each of the firings is represented by dots on the profile, and the dots are connected by lines. In this way the firing is graphically represented, and the quantity consumed during a period of a few hours can be quite accurately determined.

The firemen were picked up in New York; had been employed on board of ocean steamers, and had never been inside of the building till the experiments were commenced. They were directed to keep the water, as near as possible, a certain level, and their fires always in one condition. The boiler pressure was varied from time to time to test the comparative economy of the engine under different pressures and speeds. The watch of the firemen was 12 hours on and 12 off.

The coal used during the whole trial:

| | |
|---|---|
| Of Delaware and Hudson Canal coal.................... | 11,800 lbs. |
| Buck Mountain .................................... | 2,360 " |
| Total............................................. | 14,160 " |
| The total quantity of clinker.......................... | 645 " |
| Small coal in ashes.......... ...................... | 383 " |

## EXPLANATION OF INDICATOR CARDS.

Indicator Cards are given on page 144, taken from the steam cylinder and the pumps. The times at which they were taken correspond nearly to the times of maximum and minimum pressure; in the main at midnight the draft in the city is usually the least, and at 10 A. M. the most; and, conversely, the least load on the engine is about midnight, and the greatest from about 7 A. M. to 5 P. M., as will be seen by profile, page 145.

It will be observed on the indicator cards, two lines are heavy, and two light; the space included between the light lines represents the pressures on the lower pump *P*; that between the heavy lines, the pressure on the upper pump *P'*. The lines marked "*a a*," are the pressures on the induction pipe *E*, directly below the lower pump; and those "*c c*," the pressures on the rising main, directly above the upper pump; these lines, during the strokes of either pump, vary inconsiderably. The upper lines, "*b b*," represent the pressures on the lower pump, in its upward stroke, while "*c c*" represent the pressures with which the water is then forced in below the bucket

The lower lines "*b b*" are the pressures with which the water is forced into the upper pump *P*, while it makes its upward strokes, and "*c c*" are the pressures above its bucket at the same time.

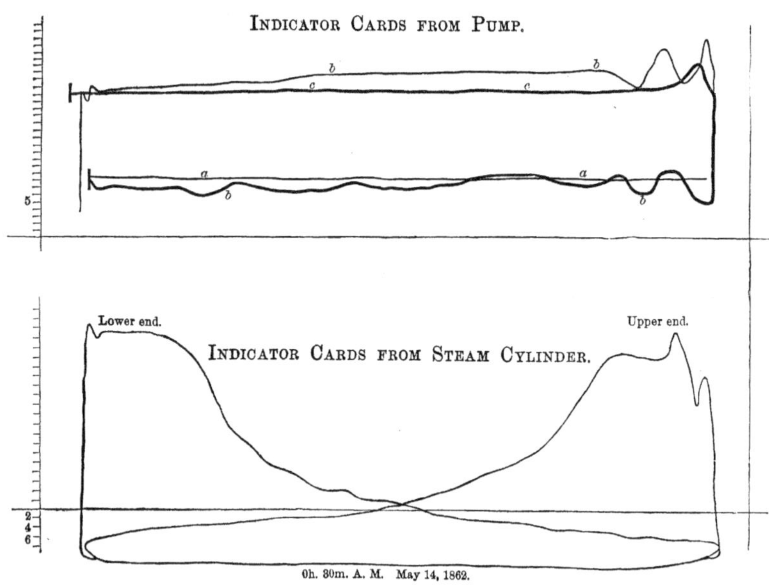


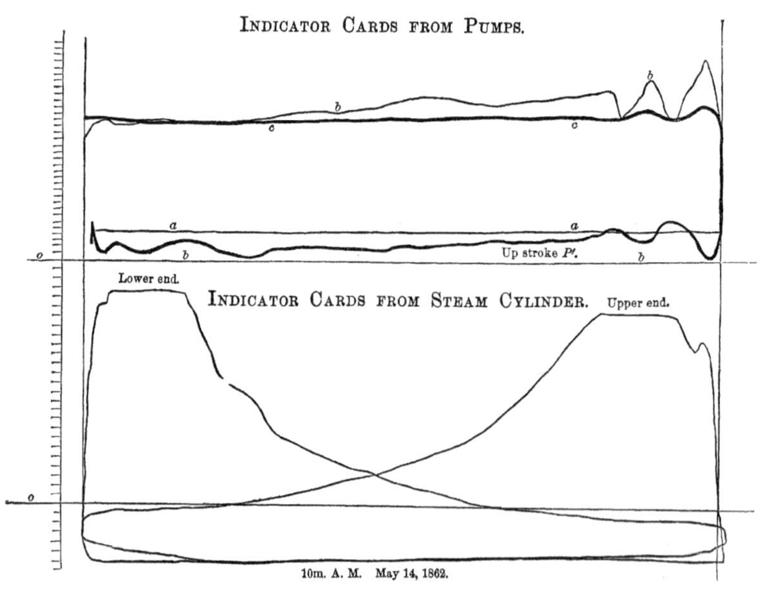

## EXPLANATION OF PROFILE OF STEAM AND WATER PRESSURES.

Indicator Cards were taken every half hour during the experiments. The average pressures on steam cards were taken in the usual way, by an average of ten ordinates; those of the pumps by the same number of ordinates, and as explained in Report and Explanation of indicator cards.

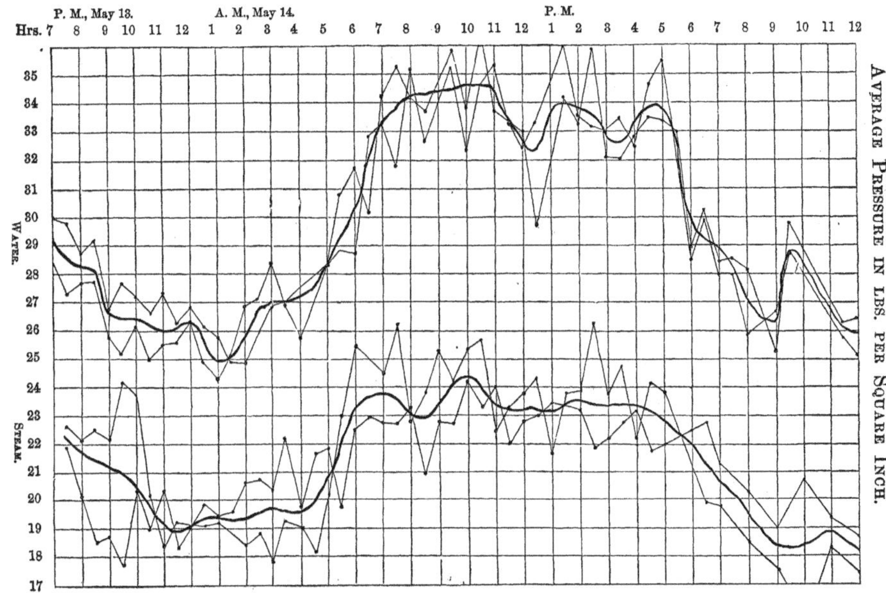


To average the pressures at each end of the cylinders, and on each of the pumps, it was thought the more correct to do this graphically, and when plotted on the same profile, they afford a comparison of the variations of power exerted and work done.

### TABLE OF AVERAGE PRESSURES ON STEAM CYLINDER AND PUMPS.

| | | PUMPS. | | STEAM CYLINDER. | |
|---|---|---|---|---|---|
| | | LOWER. | UPPER. | LOWER END. | UPPER END. |
| May 13, P. M. | 7.00 | 30. | 28.5 | | |
| | 7.30 | 29 85 | 27.3 | 22.6 | 21.9 |
| | 8.00 | 28.70 | 27.75 | 22.05 | 20.25 |
| | 8.30 | 29.20 | 27.70 | 22.55 | 18.55 |
| | 9.00 | 26.6 | 25.95 | 22.55 | 18.70 |
| | 9 30 | 27.7 | 25.25 | 24.3 | 17.6 |
| | 10.00 | 27.25 | 26.15 | 23.8 | 20.4 |
| | 10.30 | 26.6 | 25 | 20.1 | 18.7 |
| | 11.00 | 27.2 | 25.5 | 18.25 | 20.35 |
| | 11.30 | 26.4 | 25.5 | 19.3 | 18.35 |
| | 12.00 | 27 | 26.2 | | |
| May 14, A. M. | 12.30 | 26.05 | 24.95 | 19.1 | 19.9 |
| | 1.00 | 25.85 | 24.3 | 19.2 | 19.4 |

By comparison of the table with the profile, on page 145, it will be seen how the pressures are plotted. The heavy lines represent the average between the two pressures of cylinder and of pumps.

By the inspection of this profile, one may observe how the work required varies with the draft on the main during different parts of the day, and how the average pressure on the cylinder, or power exerted, varies relatively to the work required.

It will also be readily understood, that by plotting a profile of coal consumed, and averaging the pressures of steam and water, as on Profile, page 145, a much more accurate estimate may be formed of the duty of an engine, either at the steam or water end, than in the usual way, and over a much shorter period of time. The assistants in charge of the various departments continue their charge, not knowing what time or times may be selected for the test; vigilance is insured, and negligence or carelessness will leave its marks on the profiles.

TABLE OF RESULTS OF EXPERIMENTS MADE ON PROSPECT HILL PUMPING ENGINE, MAY 13TH, 14TH, 15TH, 16TH, AND 17TH, 1862.

| DATE, 1862. | | Number of Revolutions during the Hour. | AVERAGE. | | | | | | | Gross Amount of Coal consumed during the Hour. | REMARKS |
|---|---|---|---|---|---|---|---|---|---|---|---|
| | | | Mean Steam Pressure per square inch in Steam Cylinder. | Mean Water Load per square inch in Pumps. | Boiler Pressure per square inch. | Vacuum in Condenser. | STEAM CYLINDER. Initial Pressure. | STEAM CYLINDER. Cut off. Full Stroke, 1.00. | STEAM CYLINDER. Final Pressure below 0. | | |
| Day. | Hour. | 1 | 2 | 3 | 4 | 5 | 6 | 7 | 8 | 9 | |
| | | | Lbs. | Lbs. | Lbs. | In. | Lbs. | | Lbs. | Lbs. | |
| May 13.......... | 8 P. M. | 1,478 | 22.3 | 28.7 | 47.6 | 27.3 | 42 | .15 | 3.5 | 140 | 7h. Ryan commenced firing. |
| | 9 " | 1,487 | 21.7 | 27.8 | 47.3 | 27 | 39 | .15 | 5 | 145 | 9h. 3′ cleaned No. 2 fire. |
| | 10 " | 1,482 | 21.2 | 26.8 | 45.7 | 27 | 40 | .18 | 5.5 | 160 | |
| | 11 " | 1,494 | 20.4 | 26.25 | 45.0 | 27.4 | 39.5 | .14 | 6 | 140 | |
| | 12 " | 1,523 | 19.6 | 26.0 | 46-1 | 27.2 | 37 | .15 | 6.5 | 140 | 12h 3′ cleaned No. 1 fire. |
| " 14.......... | 1 A. M. | 1,578 | 19.15 | 25.5 | 45.1 | 27.2 | 35 | .15 | 6 | 125 | |
| | 2 " | 1,600 | 19.15 | 25.25 | 44.3 | 27.5 | 34 | .15 | 6 | 160 | |
| | 3 " | 1,638 | 19,25 | 26.6 | 46.5 | 27.5 | 38 | .16 | 6 | 183 | |
| | 4 " | 1,686 | 19.5 | 27.3 | 48 | 27.5 | 38.5 | .15 | 5.5 | 121 | |
| | 5 " | 1,720 | 20.1 | 27.8 | 48.5 | 27.5 | 39 | .16 | 5.5 | 160 | |
| | 6 " | 1,775 | 21.9 | 29-3 | 47 | 27.5 | 38.2 | .20 | 4 | 130 | 6h 30′ Kenny commenced firing. |
| | 7 " | 1,771 | 23.75 | 32.1 | 47.5 | 27.5 | 39 | .20 | 4 5 | 285 | |
| | 8 " | 1,680 | 24.5 | 33.9 | 45 | 27.5 | 35.5 | .17 | 3.5 | 260 | |
| | 9 " | 1,679 | 23.9 | 34.75 | 49.7 | 27.5 | 41 | .19 | 3.5 | 210 | 9h 18′ cleaned No. 1 fire. |
| | 10 " | 1,675 | 24.1 | 35.15 | 45.8 | 27.5 | 41 | .18 | 4.5 | 170 | |
| | 11 " | 1,670 | 24-1 | 34.9 | 45.6 | 27.5 | 42.5 | .18 | 3.5 | 150 | |
| | 12 " | 1,669 | 23.4 | 33.5 | 51.1 | 27.5 | 42 | .18 | 4 | 200 | |
| | 1 P. M. | 1,645 | 23.35 | 33.35 | 49 | 27.5 | 42 | .18 | 4 | 200 | |
| | 2 " | 1,663 | 23.45 | 33.75 | 48 5 | 27.5 | 41 | .19 | 3.5 | 190 | 2h 20′ cleaned No. 2 fire. |
| | 3 " | 1,663 | 23.6 | 33 65 | 44.1 | 27.4 | 36 | .19 | 3 | 220 | |
| | 4 " | 1,664 | 23.55 | 33.25 | 48.0 | 27.5 | 41.5 | .15 | 3.5 | 213 | |
| | 5 " | 1,681 | 23.15 | 33.7 | 46.2 | 27.4 | 37.5 | .18 | 3.5 | 185 | |
| | 6 " | 1,514 | 23.3 | 32.3 | 46.8 | 27.6 | 39.5 | .16 | 4.5 | 120 | 6h 25′ Ryan fires. |
| | 7 " | 1,289 | 21.4 | 29.1 | 49.7 | 27.7 | 39.2 | .15 | 4.5 | 125 | |
| | 8 " | 1,288 | 20.1 | 27.55 | 48.0 | 27.2 | 38.5 | .14 | 5.5 | 105 | |
| | 9 " | 1,285 | 19.0 | 26.5 | 48.0 | 27.2 | 37.5 | .15 | 6 | 110 | |
| | 10 " | 1,286 | 18.5 | 28.8 | 48.5 | 27.6 | 37 | .14 | 6.5 | 140 | 10h cleaned No. 1 fire. |
| | 11 " | 1,293 | 18.5 | 27.3 | 48.0 | 27.7 | 36.5 | .15 | 6 | 150 | |
| | 12 " | 1,299 | 18.4 | 25.05 | 46.5 | 27,7 | 36.5 | .15 | 6 | 120 | |
| " 15.......... | 1 A. M. | 1,279 | 18.0 | 24.45 | 46.75 | 27.7 | 35.5 | .15 | 6 | 120 | |
| | 2 " | 1,280 | 17.55 | 24 45 | 47.5 | 27.7 | 34.5 | .13 | 6 | 130 | 2h 5′ cleaned No. 2 fire. |
| | 3 " | 1,286 | 17.2 | 23.9 | 45.5 | 27.5 | 34.5 | .12 | 6 | 125 | |
| | 4 " | 1,281 | 16.95 | 24.3 | 45.5 | 27.2 | 34.5 | .13 | 6.5 | 116 | |
| | 5 " | 1,279 | 17.7 | 25.8 | 45.2 | 27.5 | 34.5 | .14 | 6.5 | 140 | |
| | 6 " | 1,286 | 19.3 | 28.6 | 46.5 | 27.7 | 38 | .17 | 5.5 | 110 | 6h 21′ Kenny fires. |
| | 7 " | 1,276 | 21.55 | 31.7 | 45.5 | 27.7 | 38 | .20 | 4.5 | 153 | |
| | 8 " | 1,258 | 22.65 | 33.55 | 37.5 | 27.7 | 32 | .25 | 4.5 | 161 | |
| | 9 " | 1,255 | 22.35 | 33.5 | 33.5 | 27.9 | 29.5 | .25 | 4 | 190 | 9h 25′ cleaned No. 1 fire. |
| | 10 " | 1,254 | 21.85 | 32.1 | 34.7 | 27.9 | 31.5 | .24 | 3.5 | 190 | |
| | 11 " | 1,256 | 21.3 | 30.5 | 34 | 27.9 | 30.5 | .23 | 3.5 | 169 | |
| | 12 " | 1,258 | 21.1 | 30.0 | 34.5 | 28 | 31 | .23 | 3.5 | 140 | 12h 32′ cleaned No. 2 fire. |
| | 1 P. M. | 1,231 | 20.95 | 30.0 | 35.5 | 28 | 31.5 | .20 | 4 | 150 | |
| | 2 " | 1,266 | 20.85 | 30.8 | 35.0 | 28 | 33 | .24 | 4.5 | 110 | |
| | 3 " | 1,255 | 20.3 | 30.5 | 35.5 | 28 | 30 | .22 | 5 | 134 | |
| | 4 " | 1,260 | 20.15 | 29.7 | 36 | 28 | 31.5 | .21 | 5 | 110 | |
| | 5 " | 1,256 | 20.25 | 29.55 | 33 | 28 | 29 | .23 | 4 | 115 | |
| | 6 " | 1,264 | 20.75 | 29.9 | 31.5 | 28 | 28.5 | .26 | 4 | 119 | 6h 40′ Ryan fires. |

| DATE, 1862. | | Number of Revolutions during the hour. | AVERAGE. | | | | | | | Gross Amount of Coal consumed during the Hour. | REMARKS. |
|---|---|---|---|---|---|---|---|---|---|---|---|
| | | | Mean Steam Pressure per square inch in Steam Cylinder. | Mean Water Load per square inch in Pumps. | Boiler Pressure per square inch. | Vacuum in Condenser. | STEAM CYLINDER. | | | | |
| | | | | | | | Initial Pressure. | Cut off. Full Stroke, 1.00. | Final Pressure below 0. | | |
| Day. | Hour. | 1 | 2 | 3 | 4 | 5 | 6 | 7 | 8 | 9 | |
| | | | Lbs. | Lbs. | Lbs. | In. | Lbs. | | Lbs. | Lbs. | |
| May 15.......... | 7 P. M. | 1,233 | 20.75 | 30.65 | 33.5 | 27.7 | 29 | .26 | 3.5 | 100 | |
| | 8 " | 1,277 | 19.9 | 27.6 | 32 | 27.5 | 25.5 | .23 | 3.5 | 96 | |
| | 9 " | 1,260 | 18.6 | 26.8 | 31.5 | 27.5 | 25.5 | .20 | 5 | 90 | 9h 38′ cleaned No. 2 fire. |
| | 10 " | 1,264 | 17.95 | 26.65 | 31.5 | 27.5 | 24.5 | .19 | 5.5 | 120 | |
| | 11 " | 1,267 | 17.8 | 26.4 | 31.5 | 27.5 | 26.5 | .18 | 5.5 | 195 | 11h 18′ cleaned No. 1 fire. |
| | 12 " | 1,266 | 17.9 | 25.95 | 32 | 27.5 | 25.5 | .22 | 5.5 | 144 | |
| " 16.......... | 1 A. M. | 1,252 | 17.7 | 25.5 | 27 | 27.6 | 21 | .27 | 5 | 141 | |
| | 2 " | 1,244 | 17.35 | 25.35 | 21.5 | 27.7 | 17 | .30 | 4.5 | 130 | |
| | 3 " | 1,241 | 17.35 | 25.35 | 21 | 27.6 | 17 | .33 | 3.5 | 100 | |
| | 4 " | 1,238 | 17.5 | 25.35 | 21 | 27.6 | 17.5 | .28 | 3.5 | 140 | |
| | 5 " | 1,239 | 17.8 | 25.9 | 20.5 | 27.7 | 17 | .30 | 4 | 130 | 6h 30′ Kenny fires. |
| | 6 " | 1,233 | 18.3 | 27.8 | 19.5 | 27.6 | 15.5 | .42 | 2.5 | 160 | |
| | 7 " | 1,229 | 19.15 | 30.4 | 20 | 27.4 | 16.5 | .38 | 1 | 164 | |
| | 8 " | 1,198 | 20.5 | 31.95 | 21 | 26.9 | 20 | .37 | 1 | 125 | 8h 55′ cleaned No. 1 fire, |
| | 9 " | 1,204 | 21.15 | 32.3 | 20 | 26.5 | 19.5 | .43 | 1 | 130 | |
| | 10 " | 1,203 | 20.7 | 31.9 | 20.5 | 26.5 | 18.5 | .38 | 1 | 150 | 10h 14′ cleaned No. 2 fire. |
| | 11 " | 1,175 | 20.4 | 31.15 | 20 | 27 | 19 | .35 | 2.5 | 204 | |
| | 12 " | 1,262 | 21.3 | 31.15 | 23 | 27.5 | 23 | .31 | 2.5 | 205 | 12h 10′ Ridgewood fireman commenced firing. |
| | 1 P. M. | 1,490 | 22.5 | 33.3 | 38 | 27.5 | 33 | .18 | 3 | 155 | |
| | 2 " | 1,480 | 22.5 | 34.15 | 49 | 27.7 | 43 | .19 | 3.5 | 120 | |
| | 3 " | 1,482 | 21.7 | 34.1 | 48.5 | 28.2 | 40.5 | .19 | 3.5 | 190 | 3h 56′ cleaned No. 1 fire. |
| | 4 " | 1,482 | 21.5 | 34.61 | 48.5 | 28.2 | 41 | .20 | 3.5 | 165 | |
| | 5 " | 1,431 | 21.3 | 35.75 | 48.5 | 27.9 | 40 | .18 | 4.5 | 190 | |
| | 6 " | 1,405 | 21.4 | 32.15 | 49 | 28 | 40 | .19 | 4.5 | 195 | 6h 40′ Ryan fires. |
| | 7 " | 1,403 | 21.35 | 32.35 | 48 | 27.9 | 40 | .20 | 4 | 180 | |
| | 8 " | 1,406 | 21.45 | 31.7 | 49 | 27.5 | 40.5 | .18 | 4 | 160 | |
| | 9 " | 1,404 | 21.5 | 30.0 | 50 | 27.5 | 41 | .17 | 4.5 | 170 | |
| | 10 " | 1,408 | 20.7 | 28.75 | 49.5 | 27.5 | 40 | .17 | 4.5 | 145 | |
| | 11 " | 1,396 | 19.4 | 27.95 | 49 | 27.5 | 38 | .15 | 5.5 | 165 | 11h 50′ cleaned No, 2 fire. |
| | 12 " | 1,356 | 18.7 | 27.15 | 47 | 27.6 | 37 | .16 | 5.5 | 180 | |
| " 17.......... | 1 A. M. | 1,267 | 18.3 | 26.7 | 47.5 | 27.6 | 34.5 | .05 | 4.5 | 120 | |
| | 2 " | 1,313 | 18.6 | 26.75 | 49.5 | 27.5 | 32.5 | .07 | 4.5 | 120 | |
| | 3 " | 1,308 | 19.15 | 26.95 | 49 | 27.5 | 31 | .12 | 4.5 | 130 | |
| | 4 " | 1,333 | 20.0 | 27.15 | 49.5 | 27.5 | 35 | .18 | 4.5 | 120 | |
| | 5 " | 1,385 | 20.75 | 28.5 | 48.5 | 27.5 | 39.5 | .13 | 4.5 | 120 | 5h 30′ cleaned No. 1 fire. |
| | 6 " | 1,382 | 20.95 | 32.8 | 47.5 | 27.1 | 39.5 | .20 | 3.5 | 144 | 6h. 30′ Ridgewood fireman commenced firing. |
| | 7 " | 1,376 | 23.3 | 35.5 | 48 | 27 | 42.25 | .21 | 2.5 | 166 | |
| | 8 " | 1,376 | 24.3 | 36.7 | 47.5 | 27 | 42.5 | .21 | 2.5 | 150 | |
| | 9 " | 1,375 | 24.0 | 37.5 | 47 | 27 | 41 | .22 | 2.5 | 150 | |

NOTE.

| | Lbs. |
|---|---|
| Used Delaware and Hudson Coal, from commencement to 10 A. M., May 16th | 10,500 |
| At 10.5 A. M., May 16th, used Buck Mountain Coal | 100 |
| Used Delaware and Hudson Canal Coal, from 10.35 A. M., Friday, to 31 min. P. M., Friday, 16th | 500 |
| Used Buck Mountain Coal, from 1.23 P. M., Friday, to 11.49 P. M., Friday | 1,900 |
| Used Delaware and Hudson Canal Coal, from 12.45 P. M., Friday, to 6.10 A. M., Saturday | 800 |
| Used Buck Mountain Coal from 6.37 A. M., Saturday, 17th, to end of trial | 360 |
| Total Coal used | 14,160 |
| Amount of Coal in Furnace No. 1 | 7,330 |
| " " " No. 2 | 6,830 |
| Clinkers during trial | 645 |
| Ashes during trial | 959 |
| 40 per cent. of which was found to be small coal | 383 |

### EXPLANATION OF TABLE.

Column 1 contains the number of revolutions during the hour, found by subtracting from the number, as shown by the counter at the hour on the line, the number at the previous hour. The number of revolutions from the commencement of the experiments, 12 M. to 7 P. M., was 11,119.

Column 2 contains the mean pressure of steam in the cylinder during the hour, and was found, first, by averaging the pressure on the cards, as taken by the indicator every half hour, then plotting these pressures on a profile, and taking the mean, as shown at page 145.

Column 3 contains the mean pressure of water load in pumps, taken from the indicator cards, and profile plotting, like that for the steam pressure, column 2.

Column 4 contains the average boiler pressure found by observation of the steam-pressure gauge in the engine room.

Column 5 contains an average of the vacuum in the condenser, taken from observations of the vacuum gauge in the engine room.

Column 6 contains the average pressures in the steam cylinder at commencement of stroke, taken from the indicator cards, plotted on profile, as at page 145.

Column 7 shows the average point at which the steam was cut off in the cylinder, taken from the indicator cards, and also plotted in profile.

Column 8 contains the average pressure in cylinder at the end of stroke, determined like that of the initial pressures, column 6.

Column 9 contains the amount of coal consumed during the hour, no deduction being made for small coal or cinders in the ashes. The amounts are determined from the profile, as in page 142. Coal used from 12 M. to 7 P. M., May 15, 1,265 pounds.

### EVAPORATION.

The evaporation—as determined from the volumes of steam in the cylinder at the end of the stroke, and referred by Pambour's formula to water at 32° Fahrenheit, and then by the English Admiralty formula to the standard of 100° Fahrenheit—was found to be 7,565 pounds of water per pound of coal.

## SPECIFICATIONS OF CAST-IRON DISTRIBUTION PIPES, AND PIPE MAINS, WITH THEIR BRANCHES, ETC.

---

1st. The pipes, branches, and all castings relating thereto, are to be delivered at such wharf or wharves in the city of Brooklyn as shall be designated by the Engineer.

2d. Every pipe below twelve inches in diameter to be nine feet in length, the pipes of twelve inches diameter and upwards to be twelve feet in length each; the diameters hereinafter specified are always to be understood as the inside diameters of the body of the pipe.

Every pipe is to have the initials of the maker's name cast distinctly upon it, and also a number signifying the order of its casting in point of date; the year (1862) to be placed on each pipe above or below said number as ($\overset{1862}{1}$, $\overset{1862}{2}$, $\overset{1862}{3}$, $\overset{1862}{4}$, &c., &c.); each size of pipe to have its own series of numbers (1, 2, 3, 4, 5, 6, &c.); each figure to be at least two to three inches in length, according to size of pipe, with a proportionate width, the weight of each pipe to be conspicuously painted on the outside or inside, before delivery, with white lead paint.

3d. The ordinary pipes shall be of the kinds usually called spigot and faucet, or socket pipes; the curves, branches, and bends, and all other special pipe castings (including such variety of branch, bevel-hub, double hub, hydrant bends, taper pipes, caps, sleeves, and any other pieces for connecting with these that may be required), shall be made according to such particular drawings and instructions as may be given by the Engineer from time to time. Every pipe shall have a bead or fillet at the spigot or small end. The faucet or hub end of the pipe to be of such form and thickness as the Engineer may direct.

4th. All pipes of twenty inches diameter and upwards to be formed so as to give a lead joint of not less than three eighths of an inch in thickness all around, and not more than seven sixteenths; those of twelve inches diameter and under, to be arranged for a joint not exceeding three eighths of an inch thickness all around, and not less than five sixteenths.

5th. The straight pipes of twelve inches diameter and upwards, shall be cast in dry sand moulds, vertically, with the hub end down, and the curved pipes in loam or sand. The smaller pipes may be cast at an angle with the horizon of not less than twelve degrees.

6th. The metal, which must be re-melted in the cupola or air-furnace, shall be made without admixture of cinder iron or other inferior metal, and shall be of such character as to make a pipe strong, tough, and of sound grain, and such as will satisfactorily bear drilling and cutting.

7th. The pipes shall be free from scoria, sand-holes, air-bubbles, cold-short cracks, and other defects or imperfections; they shall be truly cylindrical in the bore, straight in the axis of the straight pipes; and true to the required curvature or form in the axis of the other pipes; they shall be internally of the full specified diameters, and shall have their inner and outer surfaces concentric. No plugging or filling will be allowed.

8th. They shall be perfectly fettled and thoroughly cleansed; no lumps or rough places shall be left in the barrels or sockets. Great care shall be taken to have the sockets of the required size to receive the spigots, having due regard to the allowance to be made for the lead joint. No pipes will be received which are defective in joint room, whether in consequence of eccentricity of form or otherwise.

9th. The spigot ends of all the branches to have lugs or horns cast on each of such form, and in such number and place as may be directed. Both spigot and faucet ends of branches, and of all other special castings, must conform to the pipes with which they are intended to connect.

10th. All pipes of less than twelve inches in diameter to have three belts cast on each pipe outside; these belts to be not less than three and a half inches wide each, and not less than three sixteenths inch thicker than the pipe; one of these belts to touch the rise at the hub end of the pipe, one to be placed twenty-five inches from the spigot end, and the third one to be placed intermediate; the twelve-inch pipes to have but one belt in the centre of the pipe of same dimensions as above described. These belt projections will not be left sharp, but must be rounded off either way.

11th. The forms, sizes, materials, strength, uniformity, and conditions of all pipes, branches, and all other castings herein referred to, shall be subject to the inspection and approval of the Engineer of the Brooklyn Water Works, and any directions or explanations required to determine the intent and meaning of these specifications will be given by him.

The Engineer or his agent to be at liberty at all times to inspect the materials in the foundry, and the moulding and casting there. Specimens of the cast-iron shall also be delivered to him when required.

In the case of the pipes for the forty-eight-inch main, a specimen rod shall be prepared for the Inspector from the metals of each day's castings. This specimen shall be of the size and form suitable for a testing machine. The Inspector will receive and label it, and take the necessary steps to ascertain its tensile strength.

The ton used will always be understood to be the gross ton of two thousand two hundred and forty pounds (2,240lbs.)

12th. Whenever the word Engineer is used herein, it refers as well to his properly authorized agents, limited by the particular duties intrusted to them.

13th. The pipes hereinafter designated as Class A pipes, shall have at each end, about twelve inches from said ends, the Roman letter A distinctly cast upon each; those of the Class B shall have, in the same manner, the letter B cast upon each pipe.

The branches of each class shall have their proper letters distinctly cast upon each "way" of the branch; all the other castings belonging to each class shall be similarly marked.

14th. Every pipe, branch, and casting, of whatever form, shall pass a careful hammer inspection under the direction of the Engineer or his inspector; and shall be subject thereafter to a proof by water pressure of three hundred pounds to the square inch for all pipes of thirty inches diameter and under, and of two hundred and fifty pounds per square inch for all pipe mains exceeding thirty inches diameter. Each pipe, while under the required pressure, shall be rapped with a hand hammer from end to end, to discover whether any defects had been overlooked.

15th. The weights of the straight pipes of nine and twelve feet in length each, are to average closely as follows, viz.:

| Thickness of required joint for calking. | Nine feet in length over all. | | | Twelve ft. in length over all. | | Permitted deviation in weight. |
|---|---|---|---|---|---|---|
| | Diameter. | Weight per pipe. | | Weight per pipe. | | |
| Inches. | Inches. | Class A. | Class B. | Class A. | Class B. | |
| | | Pounds. | Pounds. | Pounds. | Pounds. | |
| 5-16 to 3-8 | 4 | 200 | ... | .... | .... | 4 per cent. |
| 5-16 to 3-8 | 6 | 330 | 360 | .... | .... | 4 per cent. |
| 5-16 to 3-8 | 8 | 430 | 500 | 570 | 660 | 4 per cent. |
| 5-16 to 3-8 | 12 | ... | ... | 890 | 1,080 | 4 per cent. |
| 3-8 to 7-16 | 20 | ... | ... | 2,100 | 2,500 | 4 per cent. |
| 3-8 to 7-16 | 30 | ... | ... | 3,960 | 4,890 | 3½ per cent. |
| 3-8 to 7-16 | 36 | ... | ... | 4,750 | .... | 3½ per cent. |
| 3-8 to 1-2 | 48 | ... | ... | 8,300 | .... | 2½ per cent. |

No pipes will be received that are more than the permitted percentage below the specified or required weights. For any excess above the deviations allowed, no payment will be made.

16th. The branches, and all other special castings, will conform in weight and thickness of iron to the drawings and directions to be furnished by the Engineer, governed by the following considerations, where more specific dimensions and directions are not given on the drawings.

The hub to be of the same size and thickness as the hub of the straight pipe of its size, the straight portion of the spigot ends to conform similarly and proportionately in weight and thickness to the corresponding pipes.

The flat surfaces of the branches may be one eighth thicker than the average thickness of the pipes for which the branch is intended.

The curves connecting these flat surfaces and connecting the branch ends, shall be made thicker than the ordinary thickness of the corresponding pipe, to meet the increased strain there ; but this additional thickness shall not be made (at the line or point of greatest strain) more than one half in excess of the ordinary thickness of its class of pipe, without the knowledge and consent of the Engineer.

The thickness of other special castings may be increased when advisable under the same conditions.

These branches, curved pipes, and all other special castings, shall be subjected to the same proof as the straight pipe, and the maker takes the risk of this proof.

The initials of the maker, the year, and the class-letter, shall be marked on each special casting.

17th. The pipes shall be carefully coated, inside and out, with coal pitch and oil, according to Dr. R. A. Smith's process. The coating to be applied at a proper heat and in a proper manner, before any rust sets in, and before the pipes have been subjected to the water pressure proof. The pipes to be heated immediately previous to dipping them in the pitch composition. (See the separate memorandum on the subject.)

18th. The "A" pipes are designed to be laid in positions situated fifty feet and upward above mean high water, subject, in other words, to an extreme head of one hundred and twenty feet. The "B" pipes and branches are designed to be laid in positions situated below the fifty-feet plane here referred to.

19th. The Engineer may reject, without proving, any casting which is not in conformity with the specifications or the drawings furnished.

All the pipes are to be delivered in such order and proportions of each size as may be directed by the Engineer, and each delivery of pipes is to be accompanied with such special castings as may be required.

20th. All the pipes and castings contracted for must be delivered in all respects sound and conformable to the contract. The inspection is not intended to relieve the contractor of any of his obligations in this respect, and a defective pipe or casting, which may have passed the inspector at the works, or elsewhere, will be at all times liable to rejection when discovered, until the final adjustment and completion

20

of the contract. The payments to be made are predicated on the receipt of sound castings only.

21st. The Engineer reserves the right of correcting any errors or omissions in these specifications, necessary for the proper fulfilment of their intention. The action of any such correction to date from the time that the Engineer gives due notice thereof.

ENGINEER'S OFFICE, BROOKLYN, N. Y., *July* 10, 1862.

## MEMORANDUM CONCERNING DR. SMITH'S COAL PITCH VARNISH, REFERRED TO IN THE ACCOMPANYING PIPE SPECIFICATIONS.

The above application must be made under the following conditions, which must be strictly observed to insure the permanence of the coating and the efficient protection of the pipes from rusting.

1st. Every pipe must be thoroughly dressed, and made clean and free from the earth or sand which clings to the iron in the moulds; hard brushes to be used in finishing the process, to remove the loose dust.

2d. Every pipe must, likewise, be entirely free from rust when the varnish is applied. If the pipe cannot be dipped presently after being cleansed, the surface must be oiled with linseed oil to preserve it until it is ready to be dipped; no pipe to be dipped after rust has set in.

3d. The coal tar pitch is made from coal tar, distilled until the naphtha is entirely removed and the material deodorized. In England it is distilled till the pitch is about the consistence of wax. The mixture of five or six per cent. of linseed oil is recommended by Dr. Smith. Pitch, which becomes hard and brittle when cold, will not answer for this use.

4th. Pitch of the proper quality having been obtained, it must be carefully heated in a suitable vessel to a temperature of three hundred degrees Fahrenheit, and must be maintained at not less than this temperature during the time of dipping. The material will thicken and deteriorate after a number of pipes have been dipped; fresh pitch must, therefore, be frequently added, and occasionally the vessel must be entirely emptied of its old contents and refilled with fresh pitch; the refuse will be hard and brittle like common pitch.

5th. Every pipe must attain a temperature of three hundred degrees Fahrenheit before removed from the vessel of hot pitch. It may then be slowly removed and laid upon skids to drip.

All pipes of twenty inches diameter and upward, will remain at least thirty minutes in the hot fluid to attain this temperature.

6th. The application must be made to the satisfaction of the Engineer of the Brooklyn Water Works, and the material be subject at all times to his examination, inspection, and rejection.

7th. Payment for coating the pipes will only be made on such pipes as are sound and sufficient according to the specifications, and are acceptable independent of the coating.

8th. No pipe to be dipped until the authorized inspector has examined it as to cleaning and rust, and subjected it thoroughly to the hammer proof. It may then be dipped, after which it will be passed to the hydraulic press to meet the required water proof.

The proper coating will be tough and tenacious when cold on the pipes, and not brittle or with any tendency to scale off.

9th. Where the coating of any pipe has not been properly applied, and does not give satisfaction, whether from defect in material, tools, or manipulations, it shall not be paid for ; if it scales off, or shows a tendency that way, the pipe shall be cleansed inside before it can be recoated or be receivable as an ordinary pipe.

ENGINEER'S OFFICE, BROOKLYN, N. Y., *July* 10, 1862.

# RAIN RECORD FOR FLATBUSH.

| Year. | January. | February. | March. | April. | May. | June. | July. | August. | September. | October. | November. | December. | *Spring.* | *Summer.* | *Autumn.* | *Winter.* | Year. |
|---|---|---|---|---|---|---|---|---|---|---|---|---|---|---|---|---|---|
| 1826.... | 0.47 | 1.46 | 3.66 | 2,03 | 0.06 | 8.40 | 3.77 | 7.95 | 3,09 | 8.54 | 2.95 | 2.51 | 5.75 | 20.12 | 14.58 | 4.44 | 44.89 |
| 1827.... | 2.45 | 4.50 | 2.05 | 5.65 | 3.45 | 3.68 | 3.58 | 5.47 | 3.78 | 4 10 | 6.05 | 3.51 | 11.15 | 12.73 | 13.93 | 10.46 | 48.27 |
| 1828.... | 3.02 | 3.25 | 4.50 | 4.33 | 4.80 | 2.55 | 3.73 | 1.97 | 6.85 | 0.96 | 8.73 | 0.45 | 13.63 | 8.25 | 16.54 | 6.72 | 45.14 |
| 1829.... | 4.48 | 5.64 | 4.17 | 5.36 | 3.90 | 3.38 | 2.35 | 4.84 | 3.16 | 4.41 | 4.96 | 1.89 | 13.43 | 10.57 | 12.53 | 12.01 | 48.54 |
| 1830.... | 2.87 | 2.07 | 8.08 | 3.30 | 3.77 | 4.68 | 5.49 | 2.49 | 1.33 | 4.78 | 6.69 | 7.92 | 15,15 | 12.66 | 12.80 | 12,86 | 53.47 |
| 1831.... | 4.72 | 3.46 | 4.24 | 5.98 | 2.36 | 1.48 | 6.17 | 2.55 | 3.14 | 4.47 | 2.02 | 1.93 | 12.58 | 10.20 | 19.63 | 10,11 | 42.52 |
| 1832.... | 5.55 | 2.76 | 0.83 | 3.40 | 4.65 | 0.97 | 5,31 | 5.91 | 1.91 | 3.06 | 3.63 | 5.56 | 8.88 | 12.19 | 8.60 | 13.87 | 43.54 |
| 1833.... | 3.18 | 1.00 | 3.30 | 1.87 | 5.70 | 4.87 | 3.63 | 3.68 | 2.94 | 8.52 | 2.39 | 5.60 | 10.87 | 12.18 | 13.85 | 9.78 | 44.68 |
| 1834.... | 2.22 | 2.37 | 1.62 | 3.21 | 4.88 | 7.25 | 4.54 | 0.90 | 3.05 | 2.70 | 3.83 | 2.90 | 9.71 | 12.69 | 9.58 | 7.49 | 39.47 |
| 1835.... | 2.65 | 1.21 | 4.71 | 6.22 | 1.76 | 3.25 | 5.13 | 2.84 | 1.27 | 2.11 | 3.16 | 3.71 | 12.69 | 11.22 | 6.54 | 7.57 | 38.02 |
| 1836.... | 7.51 | 4.06 | 2.50 | 4.01 | 1.38 | 8.55 | 2.50 | 2.81 | 1.66 | 2.74 | 2.75 | 3.92 | 7.89 | 13.36 | 6.15 | 15.49 | 43.89 |
| 1837.... | 1.84 | 2.83 | 5.92 | 3.23 | 4.29 | 3.45 | 2.98 | 3.15 | 1.76 | 1.01 | 1.65 | 2.55 | 13.44 | 9.58 | 4.42 | 7.22 | 34.66 |
| 1838.... | 2.22 | 1.50 | 3.42 | 3.07 | 4.54 | 4.90 | 2.19 | 1.74 | 8.22 | 3.98 | 3.91 | 1.42 | 11.03 | 8,83 | 16.11 | 5.14 | 41.11 |
| 1839.... | 1.82 | 2.52 | 1.88 | 3.64 | 5.74 | 4.84 | 2.75 | 3.91 | 3.09 | 1.99 | 3.88 | 6.84 | 11.26 | 11.50 | 8.96 | 11.18 | 42.90 |
| 1840.... | 2.00 | 2.84 | 2.58 | 4.95 | 2.82 | 1.90 | 1.10 | 3.71 | 1.84 | 4.62 | 3.33 | 4.26 | 10.35 | 6.71 | 9.79 | 9.10 | 35,95 |
| 1841.... | 6.70 | 1.53 | 4.77 | 6.77 | 3.01 | 4.56 | 2.85 | 5,50 | 1.73 | 4.50 | 5.02 | 5.18 | 14.55 | 12.91 | 11.25 | 13.41 | 52,12 |
| 1842.... | 1.34 | 4.13 | 2.08 | 4.58 | 4,63 | 3.25 | 6.17 | 5.86 | 2.82 | 4.17 | 3.50 | 4.51 | 11.29 | 15.28 | 10.49 | 9.98 | 47.04 |
| 1843.... | 1.13 | 2.79 | 4.59 | 5.21 | 1,59 | 1.75 | 1.96 | 15.76 | 2.78 | 4.87 | 3.88 | 3.88 | 11.39 | 19.47 | 11.53 | 7.80 | 50.19 |
| 1844 ... | 3.66 | 1.50 | 4.37 | 0.53 | 3.83 | 3.48 | 5.42 | 3.71 | 2.75 | 4.22 | 1.79 | 2.81 | 8.73 | 12.61 | 8.76 | 7.97 | 38.07 |
| 1845.... | 3.84 | 4.45 | 2.50 | 1.04 | 1.55 | 3.36 | 1.67 | 3.82 | 2.65 | 1.92 | 3.08 | 2.81 | 5.09 | 8,30 | 7.65 | 11.10 | 32.14 |
| 1846.... | 4.16 | 3.56 | 2.78 | 3.01 | 7.18 | 1.12 | 5.75 | 4,10 | 0.09 | 1.72 | 7.12 | 3.41 | 12.97 | 10,97 | 9.94 | 11.13 | 45.00 |
| 1847.... | 3.54 | 5.41 | 5,13 | 0.82 | 1.69 | 5.51 | 2.58 | 2.40 | 10.65 | 1.79 | 3.08 | 3.33 | 7.64 | 10.49 | 15.52 | 11.28 | 33.93 |
| 1848.... | 1.80 | 1.51 | 2.12 | 0.89 | 5.27 | 5.19 | 3.00 | 1.00 | 2.15 | 4.09 | 1.85 | 4.54 | 8.28 | 9.19 | 8.09 | 7.85 | 33.41 |
| 1849.... | 0.55 | 2.23 | 5.07 | 0.80 | 3.76 | 0.77 | 2.15 | 3.99 | 0.98 | 6.26 | 1.88 | 4.03 | 9.63 | 6.91 | 9.12 | 6.81 | 32.47 |
| 1850.... | 5.70 | 3.33 | 3.75 | 2.68 | 7.21 | 2.53 | 4.43 | 7.43 | 4.19 | 1.85 | 1.10 | 4.08 | 13.64 | 14.49 | 7.14 | 13.11 | 48.38 |
| 1851.... | 1.63 | 4.24 | 3.34 | 7.02 | 4.00 | 2.00 | 3.85 | 3.23 | 1.06 | 4.47 | 3.98 | 2.00 | 14.36 | 9.08 | 9.51 | 7.87 | 40.82 |
| 1852.... | 2.47 | 3.27 | 4.34 | 5.25 | 2.40 | 2.49 | 1.77 | 9.25 | 2.01 | 2.52 | 6.24 | 5.73 | 11.99 | 13.51 | 10.77 | 11.47 | 47.74 |
| 1853.... | 2.63 | 4.29 | 3.85 | 4.04 | 5.01 | 5.68 | 5.74 | 5.96 | 5.20 | 3.66 | 4.00 | 2.38 | 12.90 | 17.38 | 12.86 | 9.30 | 52.44 |
| 1854.... | 2.83 | 5.41 | 1.62 | 11.60 | 5.34 | 2.16 | 1.69 | 0.85 | 3.23 | 2.05 | 6.25 | 3.65 | 18.56 | 4.70 | 12.53 | 11.89 | 47.68 |
| 1855.... | 4.55 | 3.72 | 2.12 | 2.20 | 4.98 | 5.50 | 5.88 | 2.62 | 2.95 | 7.08 | 3.04 | 6.20 | 9 31 | 14.90 | 13.07 | 14,47 | 51.75 |
| 1856.... | 4.13 | 0.98 | 2.13 | 2.69 | 4.18 | 3.80 | 3.24 | 0.84 | 3.84 | 0.88 | 2.68 | 3.32 | 9.00 | 7.88 | 7.40 | 8.43 | 32.71 |
| 1857.... | 1.99 | 1.56 | 2.14 | 7.50 | 6.13 | 4.57 | 6.37 | 3.81 | 3.48 | 3.88 | 1.90 | 4.94 | 15.77 | 14.75 | 9.26 | 8.49 | 48.27 |
| 1858 ... | 2.77 | 2.24 | 1.60 | 4.86 | 4.95 | 5,34 | 5.06 | 2.77 | 2.90 | 1.92 | 4.49 | 4.33 | 11.41 | 13.17 | 9.31 | 9.34 | 43.23 |
| 1859.... | 6.46 | 4.51 | 6.93 | 5.35 | 2.42 | 5.70 | 5.22 | 5.93 | 6.80 | 1.87 | 3.99 | 3.74 | 14.70 | 16.85 | 12.66 | 14.71 | 58.92 |
| 1860.... | 1.21 | 2.47 | 1.29 | 3.65 | 2.82 | 2.73 | 2.41 | 6.75 | 3.89 | 2.71 | 5.99 | 1.73 | 7.76 | 11.89 | 12.59 | 5.41 | 37.65 |
| 1861.... | 3.88 | 2.95 | 5.14 | 4.98 | 5.49 | 3.28 | 2.11 | 2.99 | 3.27 | 4.86 | 6.31 | 0.39 | 15.61 | 8.38 | 14.44 | 7.22 | 45.65 |
| 1862.... | 2.25 | 2.22 | 2.96 | 3.56 | 3.09 | 4.17 | 3.13 | 3.78 | 3.06 | 2.73 | 4.49 | 2.58 | 9.61 | 11.08 | 10.28 | 7.05 | 38.02 |
| 1863.... | 2.41 | 2.36 | 3.13 | 3.29 | 2.15 | 2.05 | 5.62 | 2.58 | 0.95 | 1.90 | 2.12 | 4.20 | 8.57 | 10.25 | 4.97 | 8.97 | 32.76 |
| 1864.... | 1.50 | 1.10 | 2.30 | 2.35 | 1.93 | 4.45 | 2.40 | 2.82 | 4.14 | 1.44 | 3.56 | 4.01 | 6.58 | 9.67 | 9.14 | 6.61 | 32.00 |
| 1865.... | 3.34 | 3.72 | 3.96 | 4.24 | 5.63 | 3.45 | 3.54 | 2.90 | 2.21 | 5.75 | 2.87 | 4.53 | 13.83 | 9.89 | 10.83 | 11.59 | 46.14 |
| Av'ge, 40 Yrs | 3.18 | 2.93 | 3.43 | 3.98 | 3.85 | 3.80 | 3.72 | 4.09 | 3.16 | 3.52 | 3.85 | 3.68 | 11.27 | 11.67 | 10.83 | 9.67 | 42.89 |

# RECORD OF RAIN-FALL AT FORT COLUMBUS, N. Y.

| YEAR. | January. | February. | March. | April. | May. | June. | July. | August. | September. | October. | November. | December. | *Spring.* | *Summer.* | *Autumn.* | *Winter.* | YEAR. |
|---|---|---|---|---|---|---|---|---|---|---|---|---|---|---|---|---|---|
| 1836.... | 1.09 | 2.01 | 1.31 | 2.60 | 0.63 | 6.46 | 1.44 | 2.37 | 3.40 | 2.00 | 1.90 | 2.30 | 4.54 | 10.27 | 7.30 | 5.40 | 27.51 |
| 1837.... | 2.70 | 3.70 | 8.20 | 7.50 | 9.50 | 8.50 | 5.90 | 6.30 | 2.10 | 2.11 | 2.90 | 6.10 | 25.20 | 20.70 | 7.11 | 12.50 | 65.51 |
| 1838 ... | 2.93 | 3.70 | 4.10 | 2.50 | 3.99 | 3.12 | 1.83 | 4.79 | 4.96 | 3.64 | 3.10 | 2.24 | 10.59 | 9.74 | 11.70 | 8.87 | 40.90 |
| 1839.... | 0.69 | 2.05 | 2.46 | 3.35 | 8.37 | 4.94 | 1.35 | 4.92 | 3.59 | 1.45 | 2.19 | 7.61 | 14.18 | 11.21 | 7.23 | 10.35 | 42.97 |
| 1840.... | 1.84 | 1.84 | 2.95 | 2.03 | 2.39 | 2.40 | 1.80 | 4.25 | 1.84 | 4.59 | 2.90 | 1.00 | 7.37 | 8.45 | 9.33 | 4.68 | 29.83 |
| 1841.... | 5.30 | 0.80 | 2.35 | 3.93 | 3.95 | 4.65 | 4.90 | 2.50 | 2.90 | 4.40 | 3.70 | 2.70 | 10.23 | 12.05 | 11.00 | 8.80 | 42.08 |
| 1842.... | 1.07 | 2.85 | 1.25 | 3.60 | 3.60 | 3.30 | 3.80 | 2.81 | 2.10 | 4.30 | 1.80 | 3.50 | 8.45 | 9.91 | 8.20 | 7-42 | 33.98 |
| 1843.... | 1.00 | 2.31 | 2.13 | 2.14 | 2.14 | 0.76 | 1.64 | 15.26 | 3.06 | 5.91 | 2.82 | 3.34 | 6.41 | 17.66 | 11.79 | 6.65 | 42.51 |
| 1844.... | 2.66 | 1.03 | 4.50 | 0.55 | 0.55 | 2.37 | 6.00 | 2.73 | 4.50 | 4.08 | 1.73 | 2.82 | 5.60 | 11.10 | 10.31 | 6.51 | 33.52 |
| 1845.... | 4.87 | 3.22 | 3.33 | 1.22 | 1.75 | 3.70 | 1.75 | 3.21 | 2 62 | 2.50 | 3.40 | 2.51 | 6.30 | 8.66 | 8.52 | 10.60 | 34.08 |
| 1846.... | 3.92 | 3.01 | 3.82 | 4.01 | 9.70 | 1.39 | 6.01 | 3.88 | 0.48 | 1.34 | 8.36 | 2.99 | 17.53 | 11.28 | 10.18 | 9.92 | 48.91 |
| 1847.... | 4.62 | 5.74 | 8.48 | 1.53 | 2.18 | 6.78 | 1.62 | 6.93 | 12.20 | 2.13 | 6.29 | 6.35 | 12.19 | 15.33 | 20.62 | 16.71 | 64.85 |
| 1848.... | 1.75 | 1.68 | 2.23 | 1.16 | 7.28 | 4.56 | 2.64 | 1.41 | 1.87 | 6.61 | 1.59 | 4.02 | 10.67 | 8.61 | 10.07 | 7.45 | 36.80 |
| 1849.... | 0.61 | 2.26 | 4.87 | 0.62 | 3.47 | 0.78 | 1.43 | 4.63 | 1.55 | 5.63 | 1.88 | 4.01 | 8.96 | 6.84 | 9.06 | 6.88 | 31.74 |
| 1850.... | 5.57 | 2.64 | 4.64 | 2.72 | 9.20 | 3.07 | 3.92 | 7.21 | 4.71 | 3.16 | 2.33 | 5.36 | 16.56 | 14.20 | 10.20 | 13.57 | 54.53 |
| 1851.... | 1.46 | 4.50 | 1.70 | 6.94 | 4.73 | 0.90 | 4.72 | 3.47 | 1.26 | 2.95 | 4.53 | 3.72 | 13.37 | 9.09 | 8.74 | 9.68 | 40.88 |
| 1852.... | 2.92 | 3.08 | 4.43 | 4.74 | 2.24 | 2.11 | 3.25 | 6.20 | 2.29 | 2.06 | 6.0 7 | 4 45 | 11.41 | 11.56 | 10.42 | 10.45 | 43.84 |
| 1853.... | 4.14 | 4.98 | 2.03 | 3.32 | 5.80 | 4.80 | 4.40 | 5-50 | 5.49 | 3.90 | 6.80 | 1.04 | 11.15 | 14.70 | 16.19 | 10.16 | 52.20 |
| 1854.... | 2.60 | 4.00 | 0.70 | 8.80 | 7.70 | 2.20 | 1.90 | 1.03 | 1.90 | 1.80 | 3.95 | 8.60 | 17.20 | 5.13 | 7.65 | 15.20 | 45.18 |
| 1855.... | 3.30 | 3.45 | 1.20 | 2.40 | 3.85 | 3,70 | 6.00 | 2.55 | 1.30 | 5.50 | 2.40 | 8.50 | 7.45 | 12.25 | 9.20 | 15.25 | 44.15 |
| 1856.... | 5.60 | 1.25 | 3.00 | 2.65 | 3.65 | 2.90 | 2.50 | 7.60 | 3.00 | 1.70 | 2.50 | 2.30 | 9.30 | 13.00 | 7.20 | 9.15 | 38.65 |
| 1857.... | 3.20 | 0.90 | 1.45 | 5.70 | 5.10 | 4.20 | 4.80 | 4.30 | 3.05 | 4.10 | 1.00 | 2.90 | 12.25 | 13.30 | 8.15 | 7.00 | 40.70 |
| 1858.... | 1.70 | 1.20 | 0.65 | 4.50 | 5.54 | 5.95 | 4.40 | 3.95 | 3.30 | 2.00 | 4.00 | 4.80 | 10.69 | 14.30 | 9.30 | 7.70 | 41.99 |
| 1859.... | 6.20 | 4.35 | 7.50 | 6.05 | 3.90 | 6.70 | 3.80 | 5.70 | 7.10 | 1.30 | 3.65 | 4.76 | 17.45 | 16.20 | 12.05 | 15.31 | 61.01 |
| 1860.... | 2.75 | 1.94 | 1.44 | 2.90 | 3.46 | 1.11 | 3.86 | 5.13 | 3.25 | 2.21 | 5.15 | 2.49 | 7.80 | 10.10 | 10.61 | 7.18 | 35.69 |
| 1861.... | 3.80 | 1.95 | 4.15 | 3.85 | 5.62 | 3.78 | 2.70 | 5.81 | 4.14 | 4.81 | 5.52 | 1.65 | 13.62 | 12.29 | 14.47 | 7.40 | 47.78 |
| 1862.... | 9.65 | 4 40 | 4.66 | 3.40 | 3.45 | 7.30 | 2.87 | 2.55 | 1.60 | 4.80 | 3.05 | 1.65 | 11.51 | 12.72 | 9.45 | 15.70 | 49.38 |
| 1863.... | 4.55 | 5.85 | 5.50 | 5.55 | 5.35 | 1.85 | 5.50 | 3.65 | 0.90 | 2.87 | 2.55 | 4.15 | 16.40 | 11.00 | 6.32 | 14.55 | 48.27 |
| 1864.... | 2.35 | 0.95 | 3.15 | 2.45 | 4.23 | 3.50 | 1.95 | 2.60 | 5.05 | 1.80 | 4.05 | 3.45 | 9.83 | 8.05 | 10.90 | 6.75 | 35.53 |
| 1865.... | 2.90 | 4.55 | 4.65 | 3.13 | 4.85 | 2.55 | 4.35 | 1.90 | 2.25 | 3.80 | 3.90 | 4.03 | 12.63 | 8.80 | 9.95 | 11.48 | 42.86 |
| Average. | 3.26 | 2.87 | 3.43 | 3.53 | 4.61 | 3.67 | 3.43 | 4.50 | 3.26 | 3.32 | 3.53 | 3.84 | 11.56 | 11.62 | 10.11 | 9.97 | 43.26 |

# RECORD OF RAIN-FALL AT FORT HAMILTON, N. Y.

| YEAR. | January. | February. | March. | April. | May. | June. | July. | August. | September. | October. | November. | December. | *Spring.* | *Summer.* | *Autumn.* | *Winter.* | YEAR. |
|---|---|---|---|---|---|---|---|---|---|---|---|---|---|---|---|---|---|
| 1843.... | 0.70 | 1.90 | 4.20 | 5.70 | 1.32 | 1.67 | 1.88 | 14.05 | 3.75 | 4.10 | 1.70 | 3.21 | 11.22 | 17.60 | 9.55 | 5.81 | 44.18 |
| 1844.... | 2.97 | 0.93 | 4.33 | 0.52 | 2.80 | 2.32 | 4.87 | 1,72 | 4.30 | 3.78 | 2.05 | 2.05 | 7.65 | 8.91 | 10.13 | 5.95 | 32,64 |
| 1845.... | 4.26 | 5.70 | 2.21 | 1.00 | 1.47 | 3.60 | 1.95 | 4.48 | 2.28 | 1.76 | 2.95 | 3.45 | 4.68 | 10.03 | 6.99 | 13.41 | 35.11 |
| 1846.... | 5.15 | 5.55 | 2.82 | 2.40 | 9.84 | 2.14 | 7.09 | 4.34 | 0.20 | 1 58 | 8.90 | 2.68 | 15.06 | 13.57 | 10.68 | 13.38 | 52.69 |
| 1847.... | 5.04 | 7,05 | 5.04 | 0.71 | 1.63 | 8.37 | 4.32 | 3.30 | 10.60 | 4.25 | 5.20 | 7.18 | 7.38 | 15.99 | 20.05 | 19.27 | 62.69 |
| 1848.... | 2.20 | 1.17 | 1.84 | 1.75 | 9.32 | 5.19 | 1.80 | 0.94 | 1.65 | 2,95 | 1.82 | 2.50 | 12.91 | 7.93 | 6.42 | 5.87 | 33.13 |
| 1849.... | 0.44 | 2.12 | 5.11 | 0.62 | 3.75 | 0.82 | 2.76 | 3.01 | 2.84 | 4.50 | 1.45 | 2.43 | 9.48 | 6.59 | 8.79 | 4.99 | 29.85 |
| 1850.... | 4.69 | 1.74 | 5,55 | 1,44 | 6.74 | 2.11 | 3.68 | 6.16 | 4.20 | 1.83 | 1.98 | 5.77 | 13.73 | 11.95 | 8.01 | 12.20 | 45.89 |
| 1851.... | 2.39 | 4.44 | 3.88 | 5.32 | 2.88 | 1.68 | 3.29 | 2.64 | 1.31 | 2.99 | 4.39 | 2.66 | 12.08 | 7.61 | 8.69 | 9.49 | 37.87 |
| 1852.... | 1,94 | 2.27 | 4.11 | 6.06 | 3.20 | 2.80 | 2.64 | 4.46 | 1.99 | 2.25 | 4.14 | 3.80 | 13.37 | 9.90 | 8.38 | 8.01 | 39.66 |
| 1853.... | 4.58 | 7.24 | 5,79 | 4.78 | 6.02 | 4,42 | 5.83 | 5,23 | 3.75 | 2.73 | 3.71 | 1.16 | 16,59 | 15.48 | 10.19 | 12.98 | 55.24 |
| 1854.... | 3.03 | 4.45 | 0,65 | 9.58 | 6.09 | 4.25 | 2.23 | 0.26 | 3.50 | 1.19 | 5.70 | 3.35 | 16.32 | 6.74 | 10.39 | 10,83 | 44.28 |
| 1855.... | 3.71 | 4.60 | 1.41 | 2.45 | 3.91 | 5.06 | 2.34 | 2.99 | 0.95 | 4.70 | 3.15 | 5.19 | 7.77 | 10.39 | 8.80 | 13.50 | 40.46 |
| 1856.... | 3.65 | 0,38 | 1.83 | 4.31 | 3.42 | 3.79 | 2.72 | 5,50 | 4.35 | 0.46 | 2.91 | 2.95 | 9.56 | 12.01 | 7.72 | 6.98 | 36.27 |
| 1857.... | 2.79 | 1.62 | 1.79 | 6.68 | 5.03 | 4.60 | 5.48 | 3.21 | 3.18 | 2.87 | 1.09 | 3.87 | 13.50 | 13,29 | 7.14 | 8.28 | 42.21 |
| 1858.... | 2.61 | 3.54 | 0.68 | 4.50 | 4.58 | 3.30 | 3.87 | 2.40 | 2.80 | 1.23 | 3.58 | 2.96 | 9.76 | 9.57 | 7.61 | 9.11 | 36.05 |
| 1859.... | 5.47 | 3.60 | 5.18 | 4.21 | 1.70 | 5.17 | 3.96 | 5.70 | 1.91 | 0.89 | 2.78 | 2.34 | 11.09 | 14·83 | 5.58 | 11.41 | 42.91 |
| 1860.... | 1.77 | 1.45 | 0,86 | 2.13 | 3.61 | 1,53 | 1.78 | 6.39 | 3.37 | 1.32 | 4.20 | 1.58 | 6.60 | 9.70 | 8.89 | 4.80 | 29.99 |
| 1861.... | 2.02 | 1.58 | 2.41 | 3.00 | 3.80 | 1.43 | 1.15 | 2.55 | 1.95 | 1.68 | 4.41 | 0.65 | 9.21 | 5.13 | 8.04 | 4.25 | 26.63 |
| 1862.... | 1.75 | 1.72 | 2.55 | 1.35 | 2.65 | 4.57 | 5.75 | 0.80 | 1.21 | 2.46 | 2.92 | 0,34 | 6.55 | 11.12 | 6.59 | 3.81 | 28.07 |
| 1863.... | 1,25 | 2.18 | 2.29 | 4.83 | 2.26 | 0.78 | 3.03 | 1.96 | 0.35 | 1.40 | 0.89 | 2.10 | 9.38 | 5.77 | 2.64 | 5.53 | 23.32 |
| 1864.... | 2.16 | 0.31 | 1.15 | 1.16 | 5.21 | 2.07 | 1.49 | 2.62 | 4.43 | 2.15 | 3.82 | 4,41 | 7.52 | 6.18 | 10.40 | 6.88 | 30.98 |
| 1865.... | 3.57, | 3.64 | 4.37 | 2.87 | 2.98 | 3.02 | 2.73 | 2.54 | 2.05 | 2.93 | 2.54 | 4.69 | 10.22 | 8.29 | 7.52 | 11.90 | 37.93 |
| Average. | 2.96 | 3.01 | 3.05 | 3.36 | 4.10 | 3.25 | 3.33 | 3.79 | 2.92 | 2.43 | 3.31 | 3.10 | 10.51 | 10.37 | 8.66 | 9.07 | 38.61 |

# STATISTICS OF THE BROOKLYN WATER SUPPLY.

| Years. | Average Daily Consumption of Water in cubic feet. | Number of miles Water Pipe laid. | Average number of Taps in use. | Pumping Expenses. | Other Expenses: Administration of the works, &c. | Total Expenditures. | Annual Receipts from Water Rates. |
|---|---|---|---|---|---|---|---|
| | | | | *Exclusive of interest on cost of Works.* | | | |
| 1859........ | ...... | 123.$^{916}$ | ...... | ........ | ........ | ........ | ........ |
| 1860........ | 440,226 | 136.$^{598}$ | 9,302 | $19,624 28 | $44,882 32 | $64,506 60 | $256,400 49 |
| 1861........ | 543,332 | 145.$^{552}$ | 12,856 | 20,047 33 | 50,841 69 | 70,889 02 | 239,355 52 |
| 1862........ | 671,248 | 157.$^{482}$ | 15,105 | 25,386 11 | 46,610 75 | 71,996 86 | 303,295 93 |
| 1863........ | 867,692 | 165.$^{622}$ | 17,145 | 40,406 35 | 54,620 64 | 95,026 99 | 362,749 80 |
| 1864........ | 1,060,480 | 171.$^{742}$ | 18,935 | 67,579 81 | 61,640 42 | 129,220 93 | 386,416 08 |
| 1865. ...... | 1,234,323 | 176.$^{704}$ | 20,382 | 64,875 89 | 72,702 39 | 137,578 28 | 419,106 32 |
| 1866........ | 1,457,824 | 183.$^{798}$ | 22,244 | 87,696 50 | 79,804 40 | 167,500 90 | 462,619 04 |

# POPULATION, ASSESSED VALUE OF PERSONAL AND REAL ESTATE, AND TAXES OF THE PRESENT CITY OF BROOKLYN.

| | Years. | Population. | Increase in five years. | Assessed Value of Personal and Real Estate. | Tax Levied. |
|---|---|---|---|---|---|
| By State Census........................................ | 1835* | 27,854 | ...... | 29,349,960 | ........ |
| U. S. Census.. ..................................... | 1840 | 42,622 | 53 per cent. | 28,645,499 | ........ |
| State Census... ..................................... | 1845 | 72,769 | 74 " " | 28,076,591 | ........ |
| U. S. Census........................................ | 1850 | 130,757 | 80 " " | 41,916,539 | ........ |
| State Census........................................ | 1855† | 205,250 | 57 " " | 94,791,215 | 1,532,692 |
| U. S. Census................. ...................... | 1860 | 266,714 | 30 " " | 103,680,566 | 1,969,794 |
| State Census.......................................... | 1865 | 296,112‡ | 11 " " | 123,427,840 | 3,795,985 |

Number of dwellings, 1860, 30,578. Productive establishments, 911.

* First charter granted to Brooklyn.

† Second charter granted, incorporating Williamsburgh and Bushwick with Brooklyn.

‡ The last census, if properly taken, would have given a much larger population. At the present time the population cannot be far from 350,000.

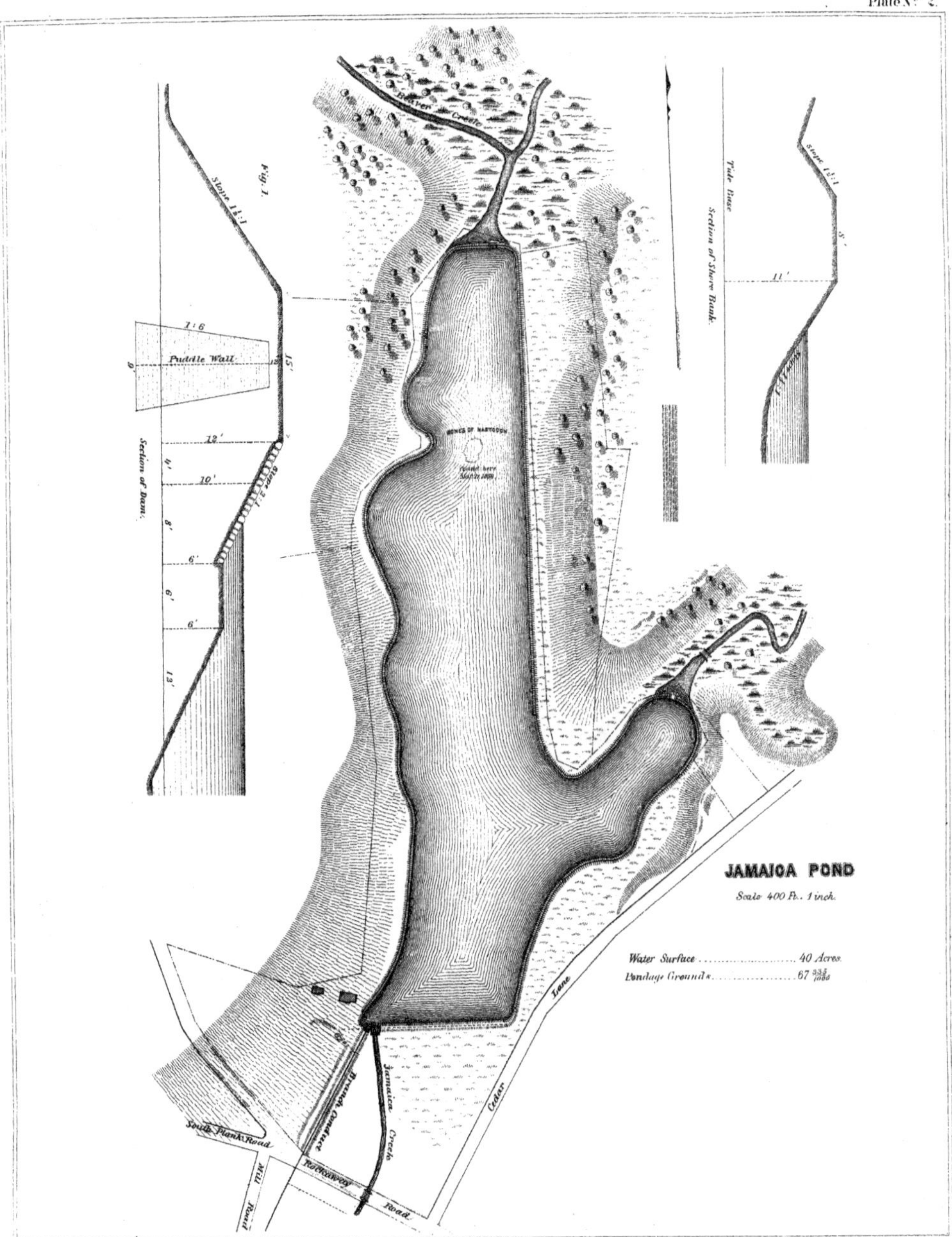


D. Van Nostrand, Publisher 192 Broadway, N.Y.

J. Bien Lith. N.Y.

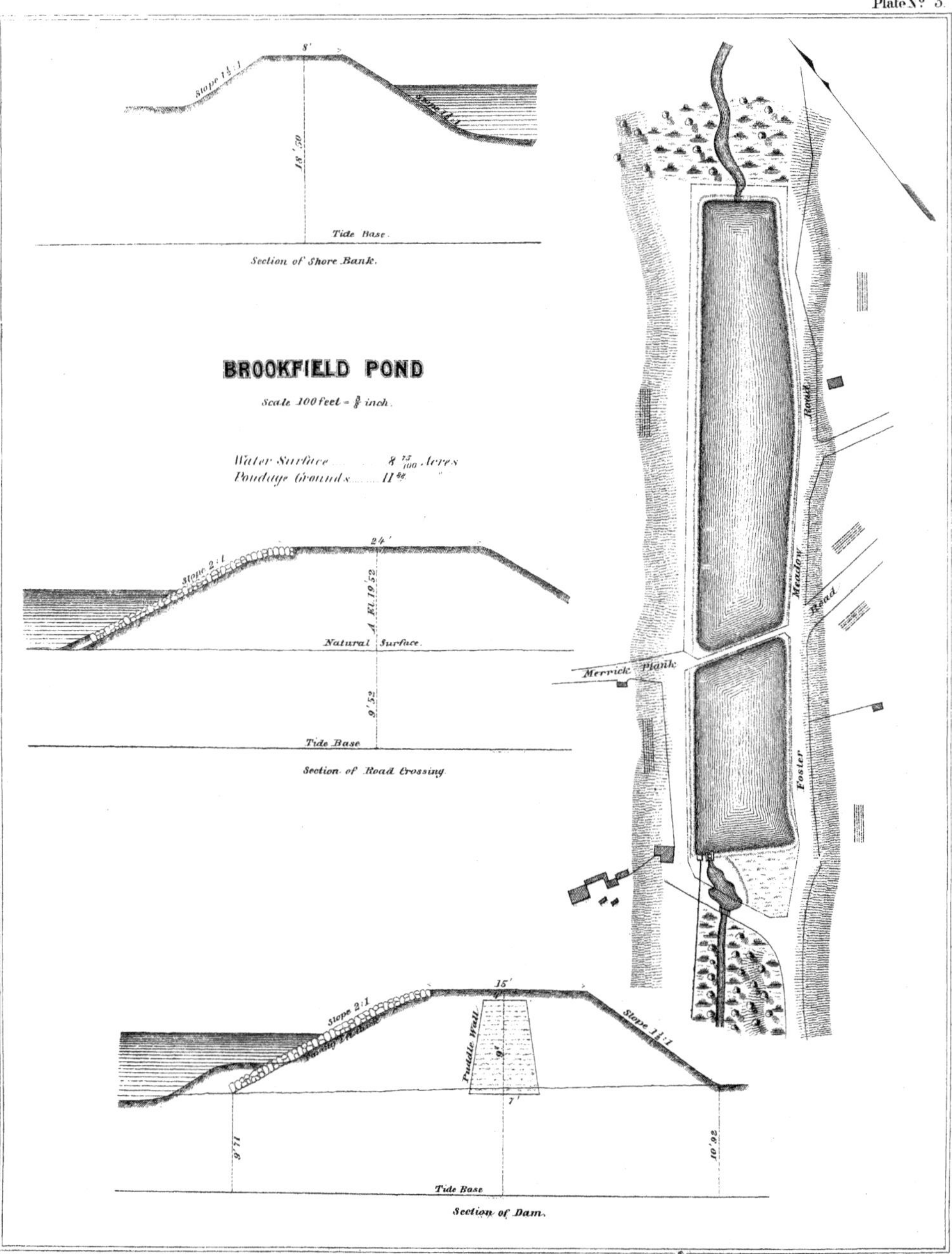


D. Van Nostrand, Publisher 192 Broadway, N.Y.

J. Bien Lith. N.Y.

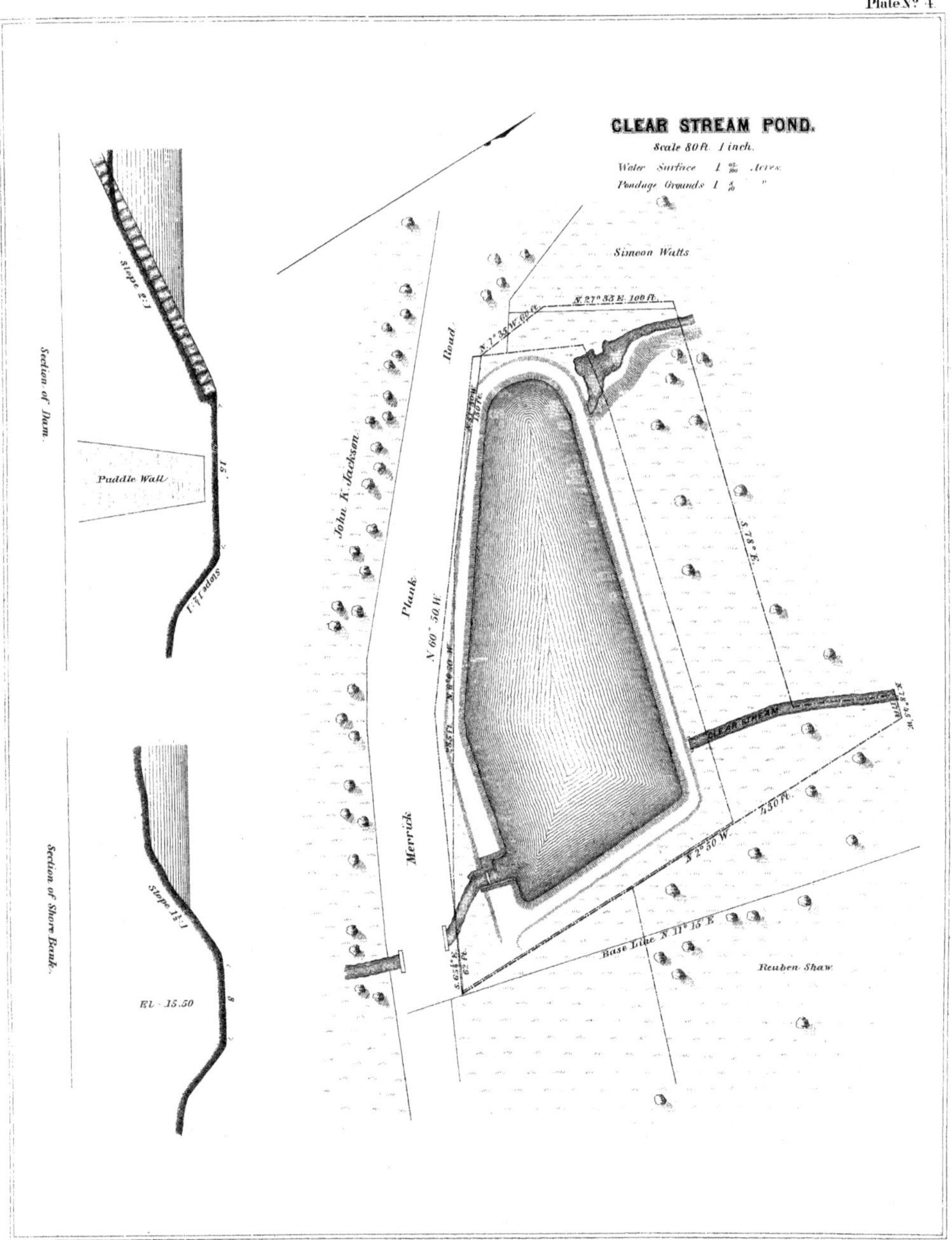


D. Van Nostrand, Publisher 192 Broadway, N.Y.
J. Bien Lith. N.Y.

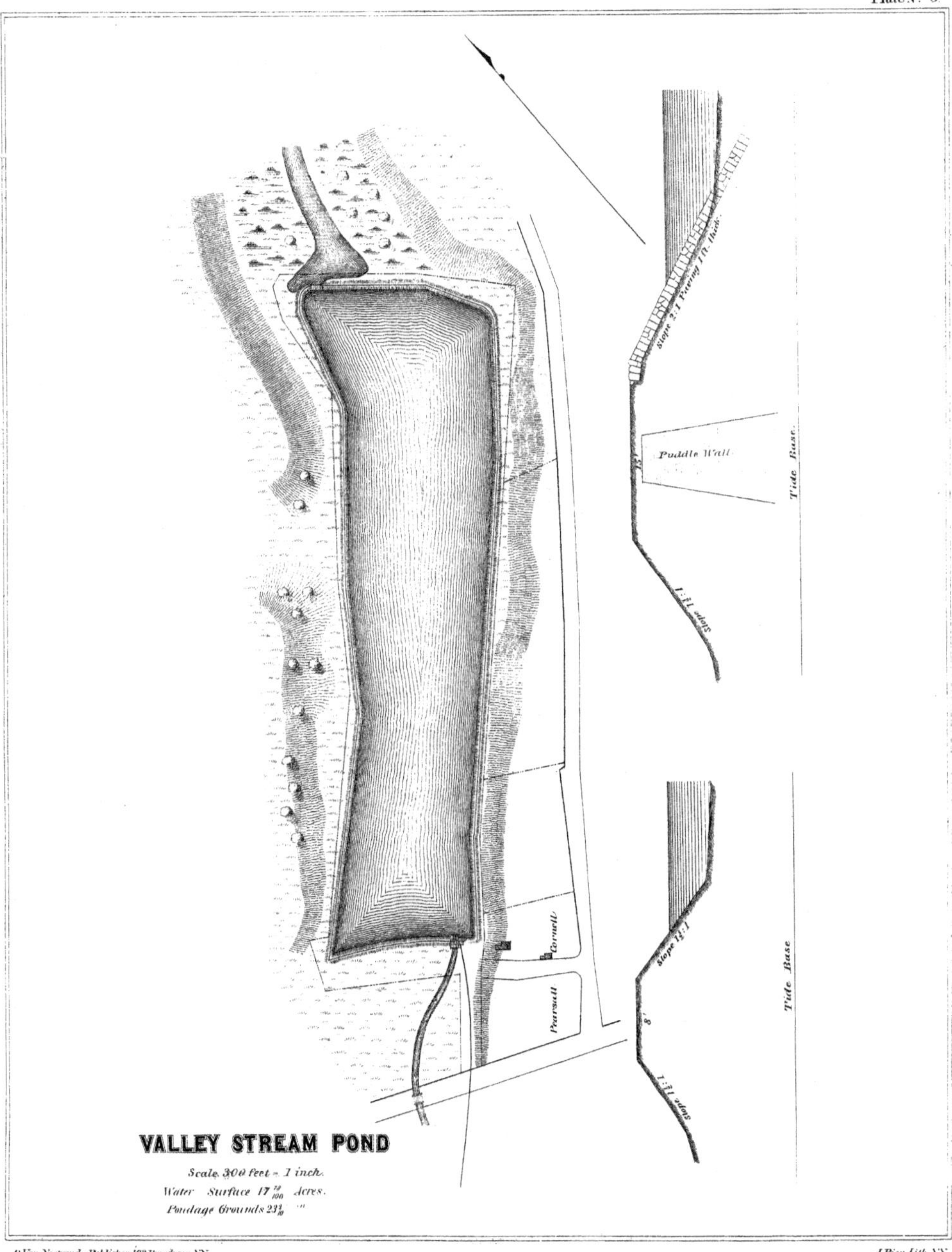


**VALLEY STREAM POND**

*Scale 300 feet = 1 inch.*
*Water Surface $17\frac{79}{100}$ Acres.*
*Pondage Grounds $23\frac{1}{10}$ "*

D. Van Nostrand, Publisher 192 Broadway. N.Y. J. Bien Lith. N.Y.

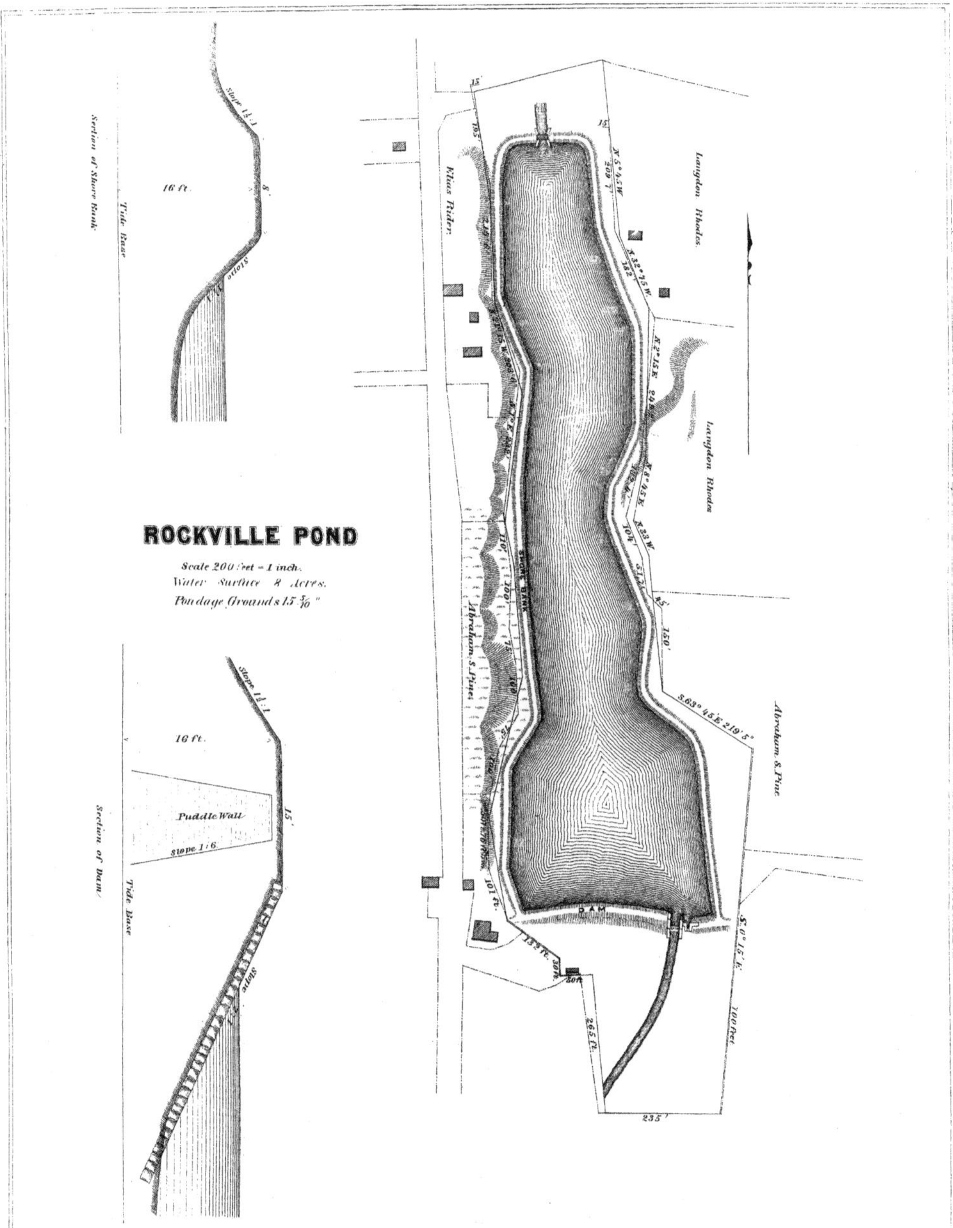


D. Van Nostrand, Publisher 192 Broadway N.Y.

F. Bien Lith. N.Y.

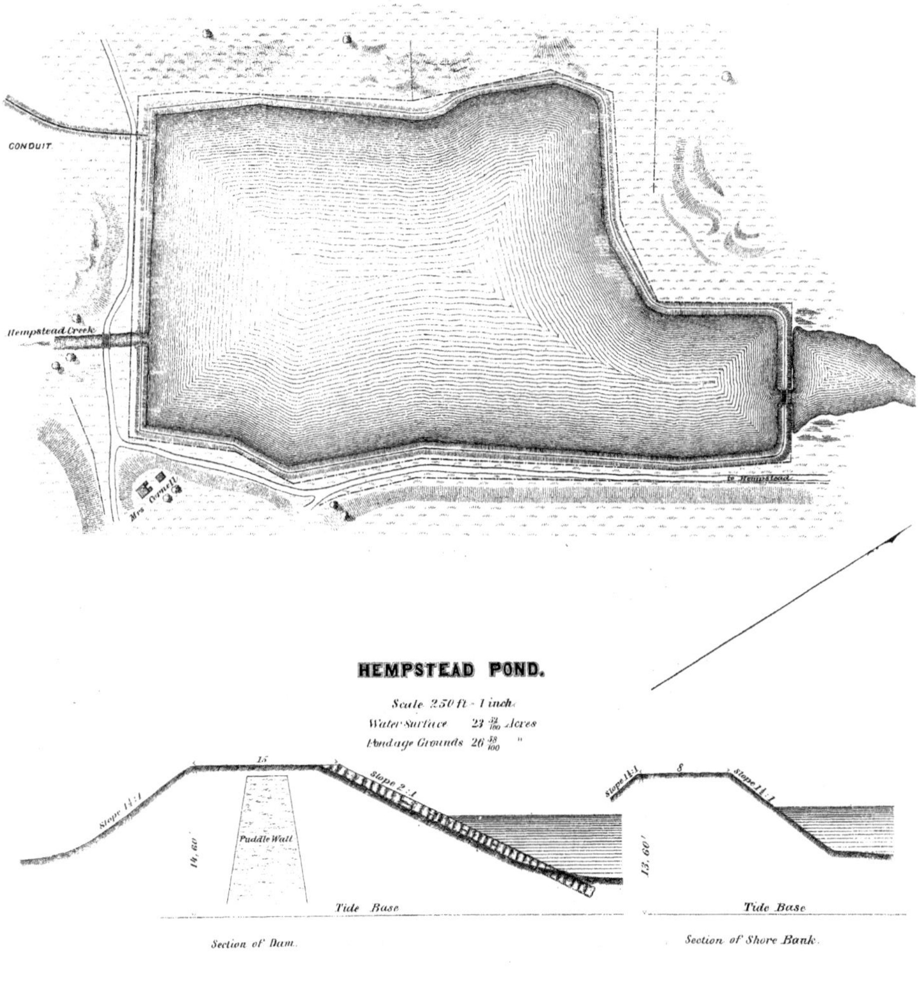

CONDUIT

Hempstead Creek

Mrs Cornell

to Hempstead

# HEMPSTEAD POND.

*Scale 250 ft = 1 inch.*

*Water Surface 23 $\frac{32}{100}$ Acres*

*Pondage Grounds 26 $\frac{38}{100}$ "*

15

Slope 1½:1

Slope 2:1

Puddle Wall

14.60

Tide Base

8

Slope 1½:1

Slope 1½:1

13.60

Tide Base

*Section of Dam.*

*Section of Shore Bank.*

D. Van Nostrand, Publisher 192 Broadway, N.Y.

J. Bien Lith. N.Y.

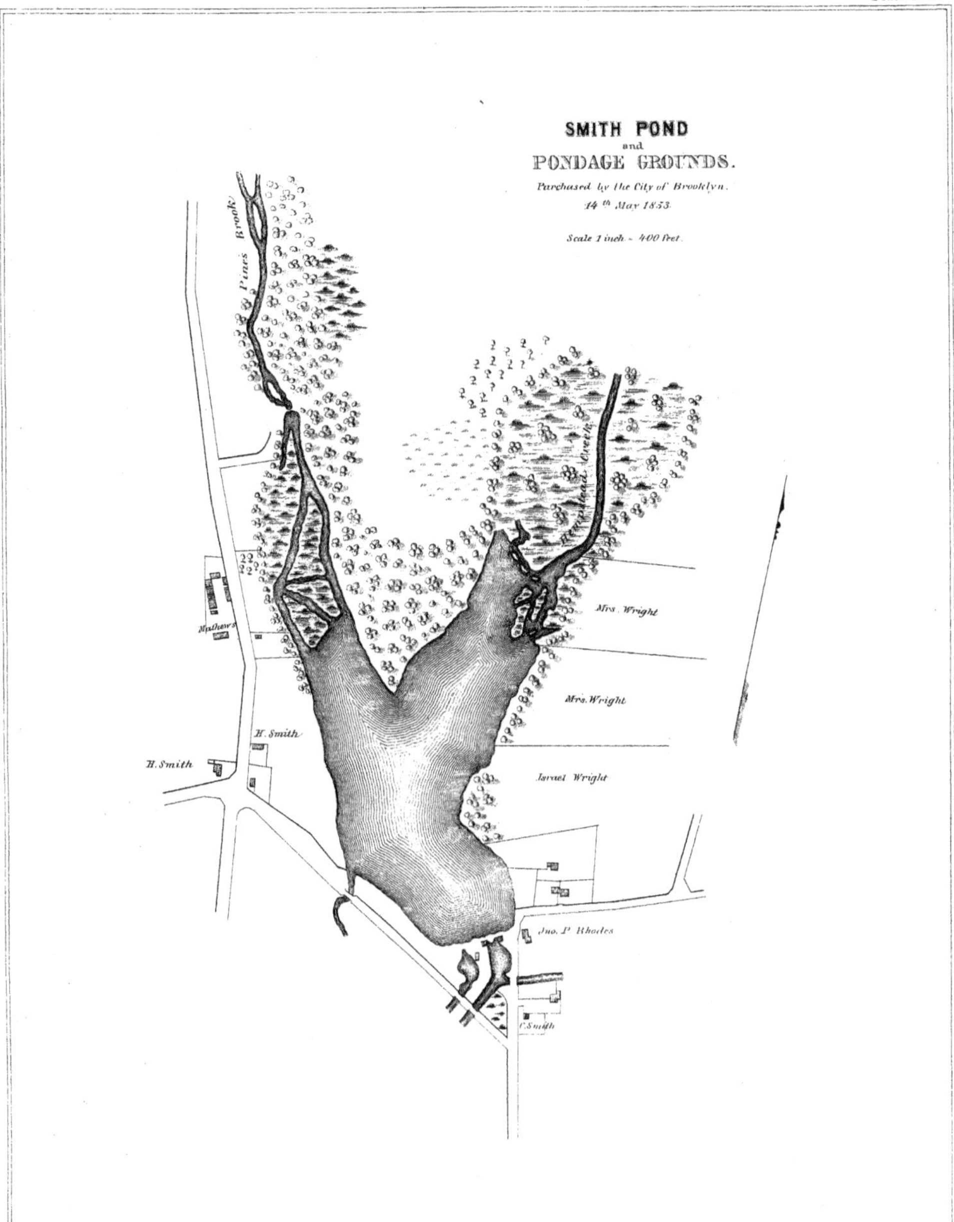


D. Van Nostrand, Publisher 192 Broadway, N.Y.

J. Bien Lith. N.Y.

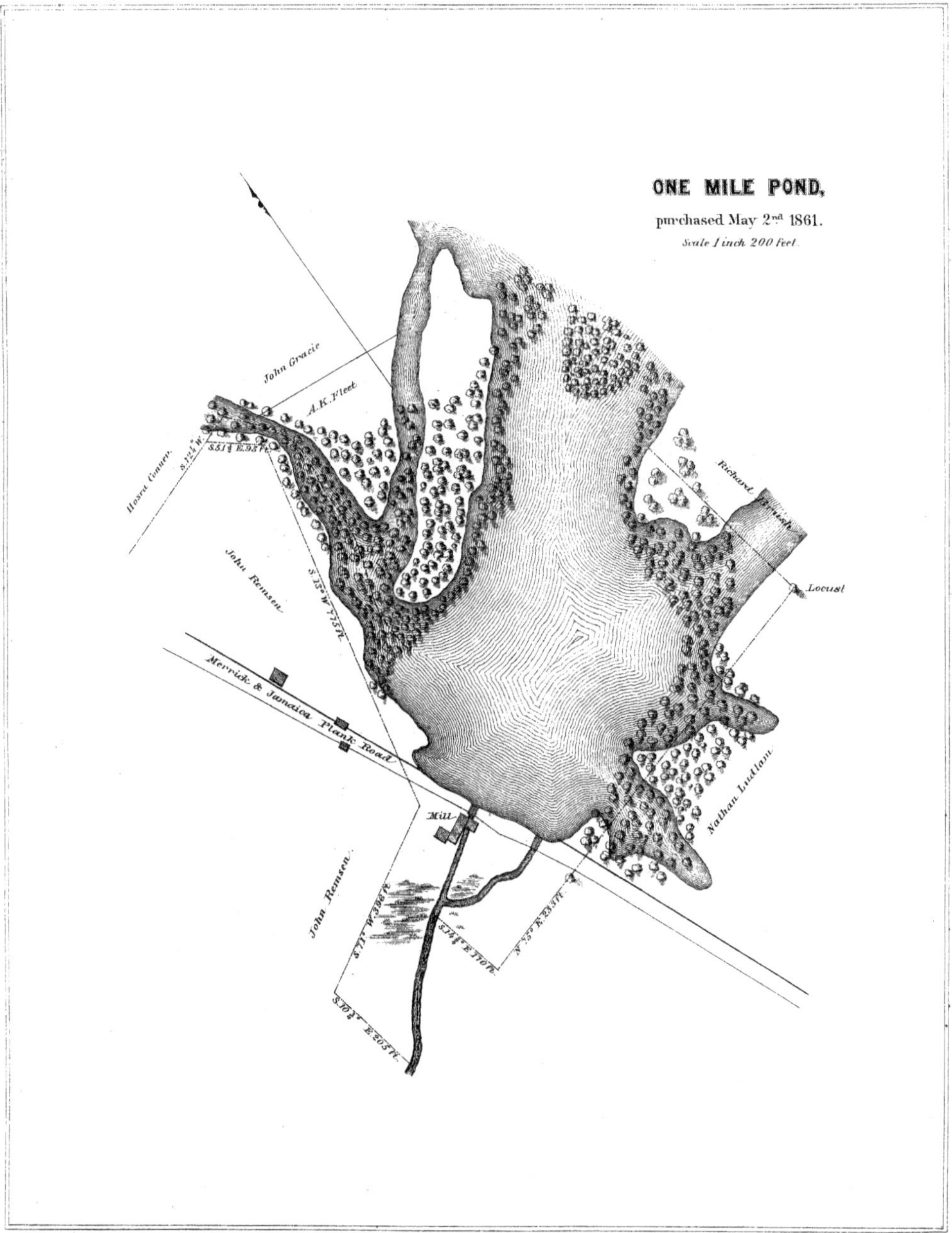


D. Van Nostrand, Publisher 192 Broadway, N.Y.

J. Bien Lith. N.Y.

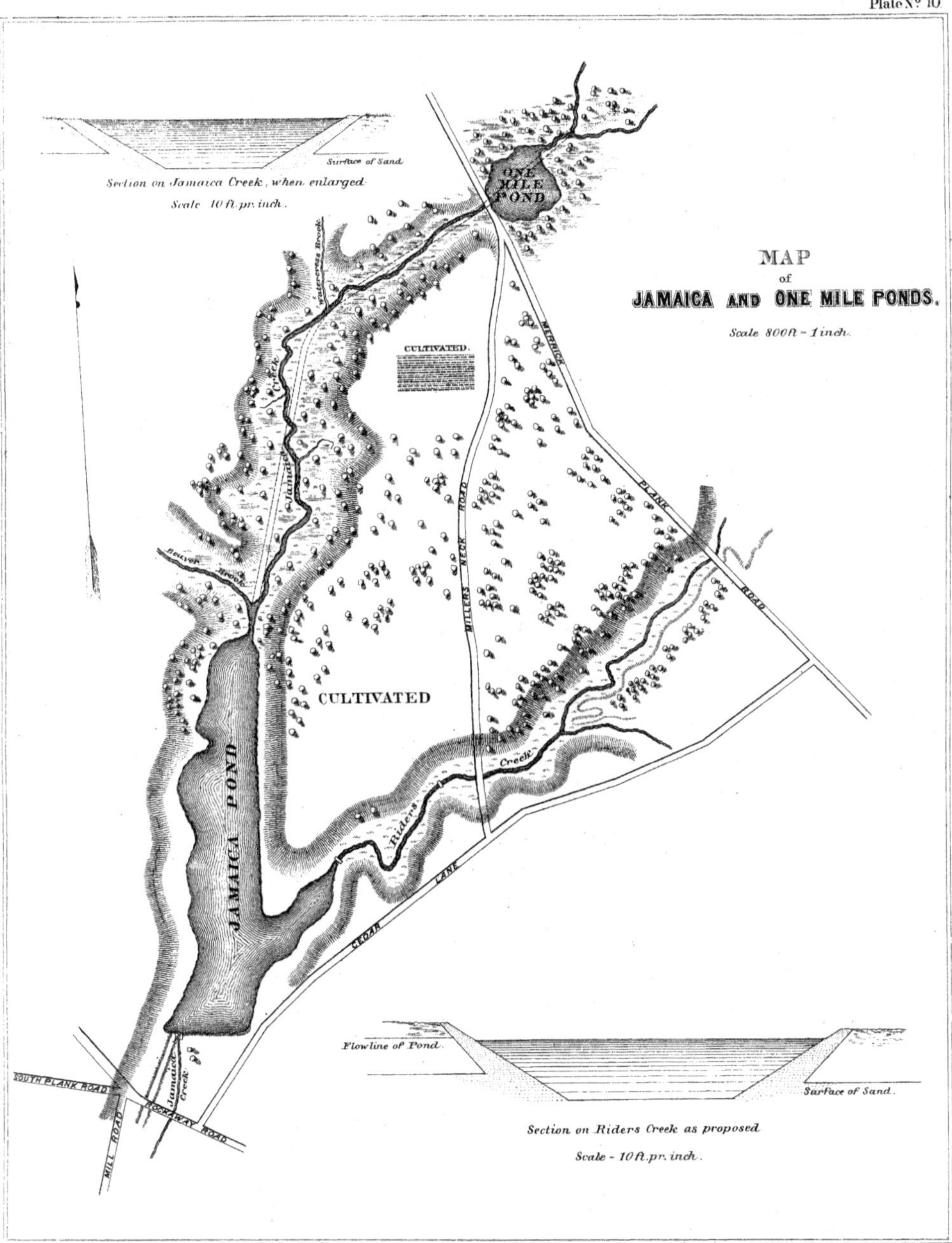


D. Van Nostrand, Publisher 192 Broadway, N.Y.
J. Bien Lith. N.Y.

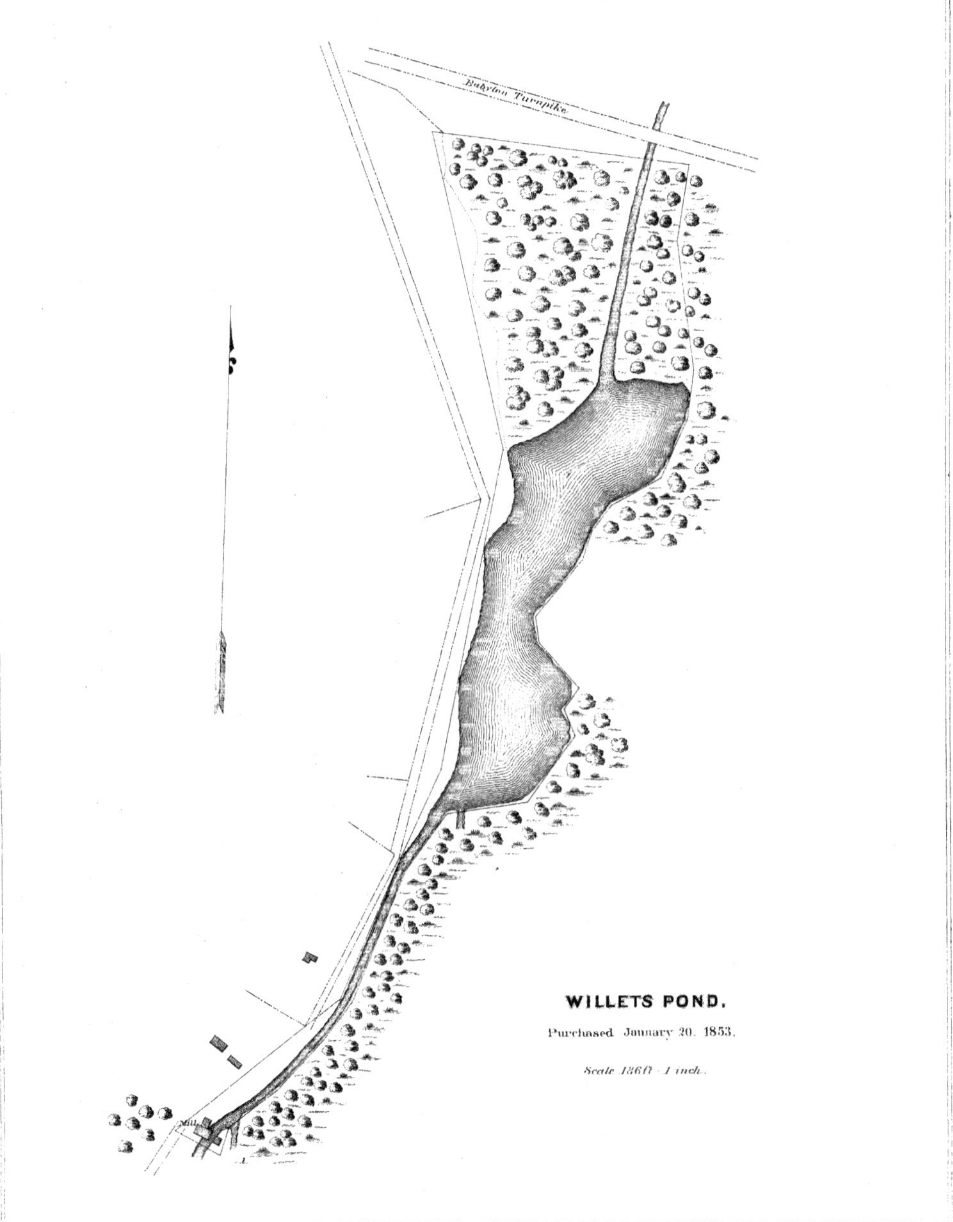


D. Van Nostrand, Publisher 192 Broadway, N.Y.

J. Bien Lith. N.Y.

Plate No. 12.

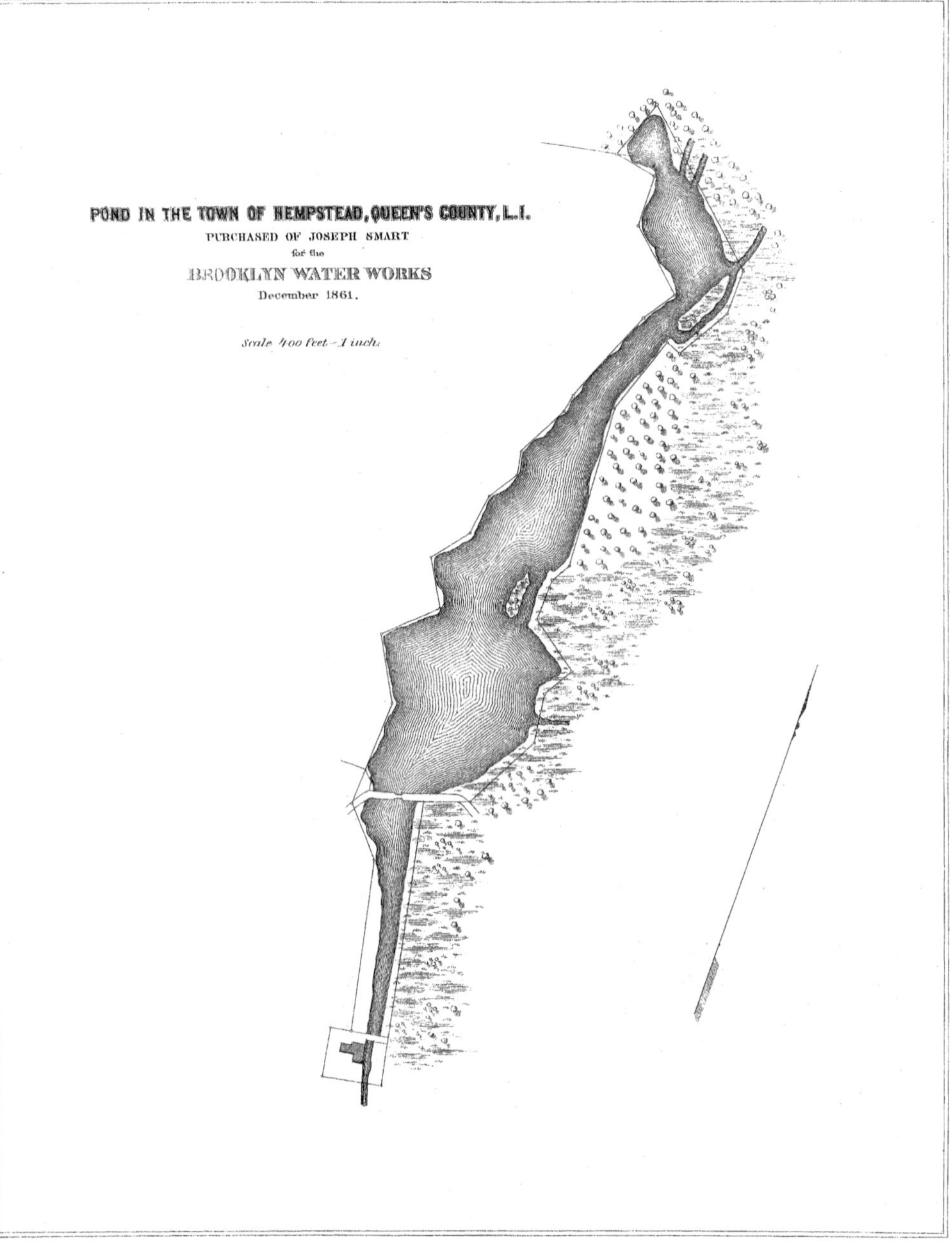


D. Van Nostrand, Publisher 192 Broadway, N.Y.

J. Bien Lith. N.Y.

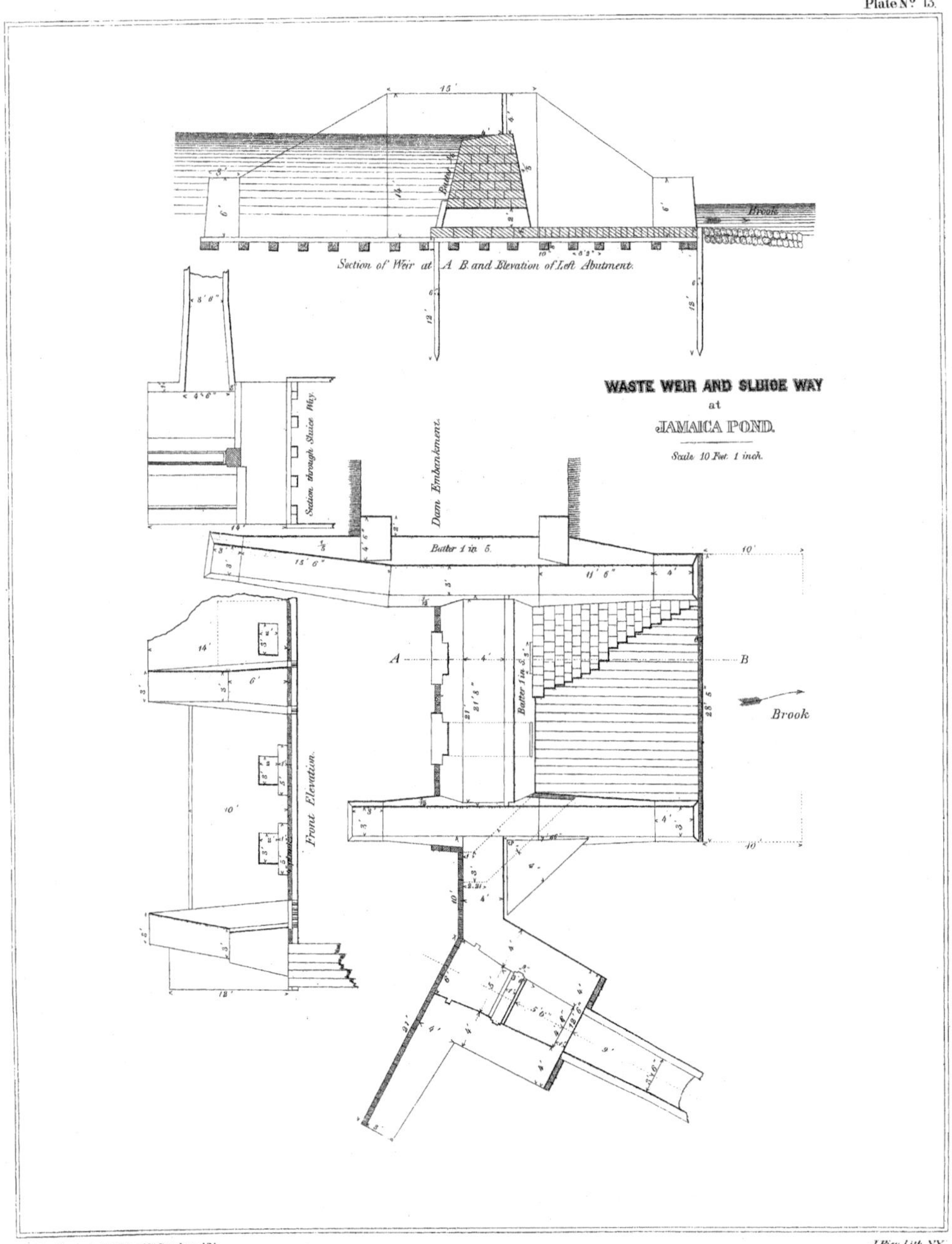


D. Van Nostrand, Publisher 192 Broadway N.Y.

J. Bien Lith. N.Y.

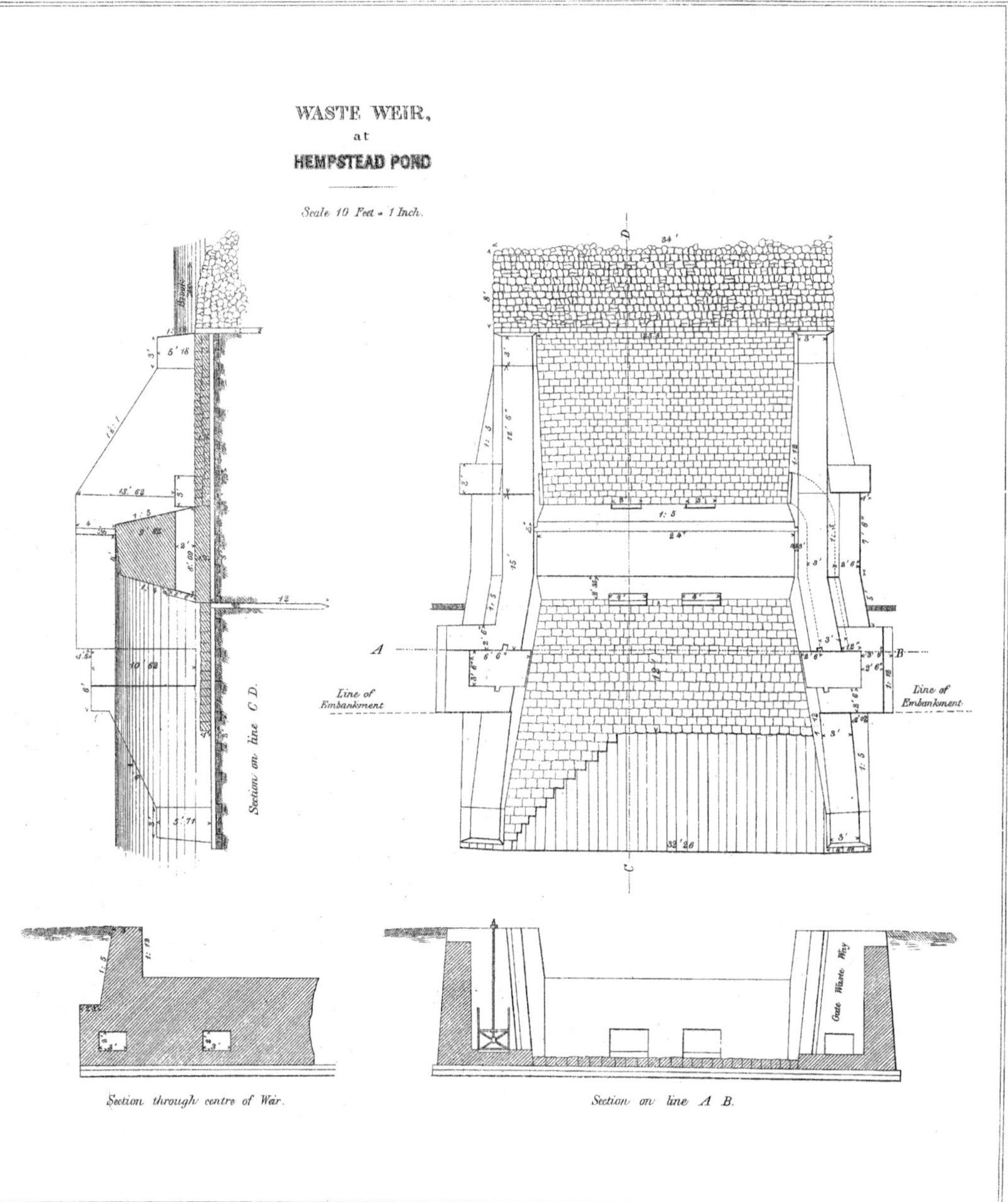


D. Van Nostrand Publisher, 192 Broadway, N.Y.

J. Bien, Lith. N.Y.

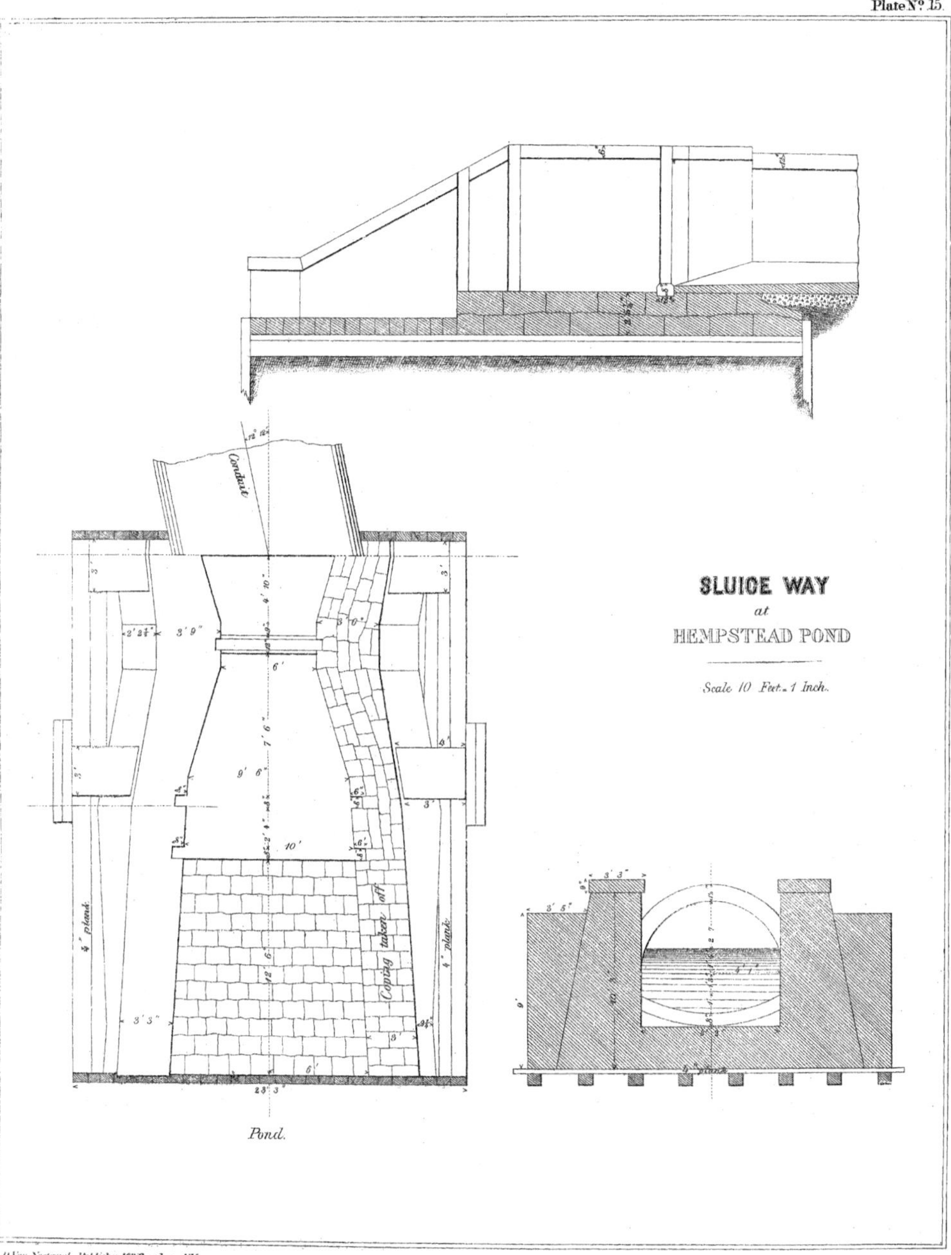


D. Van Nostrand, Publisher 192 Broadway, N.Y.
J. Bien Lith. N.Y.

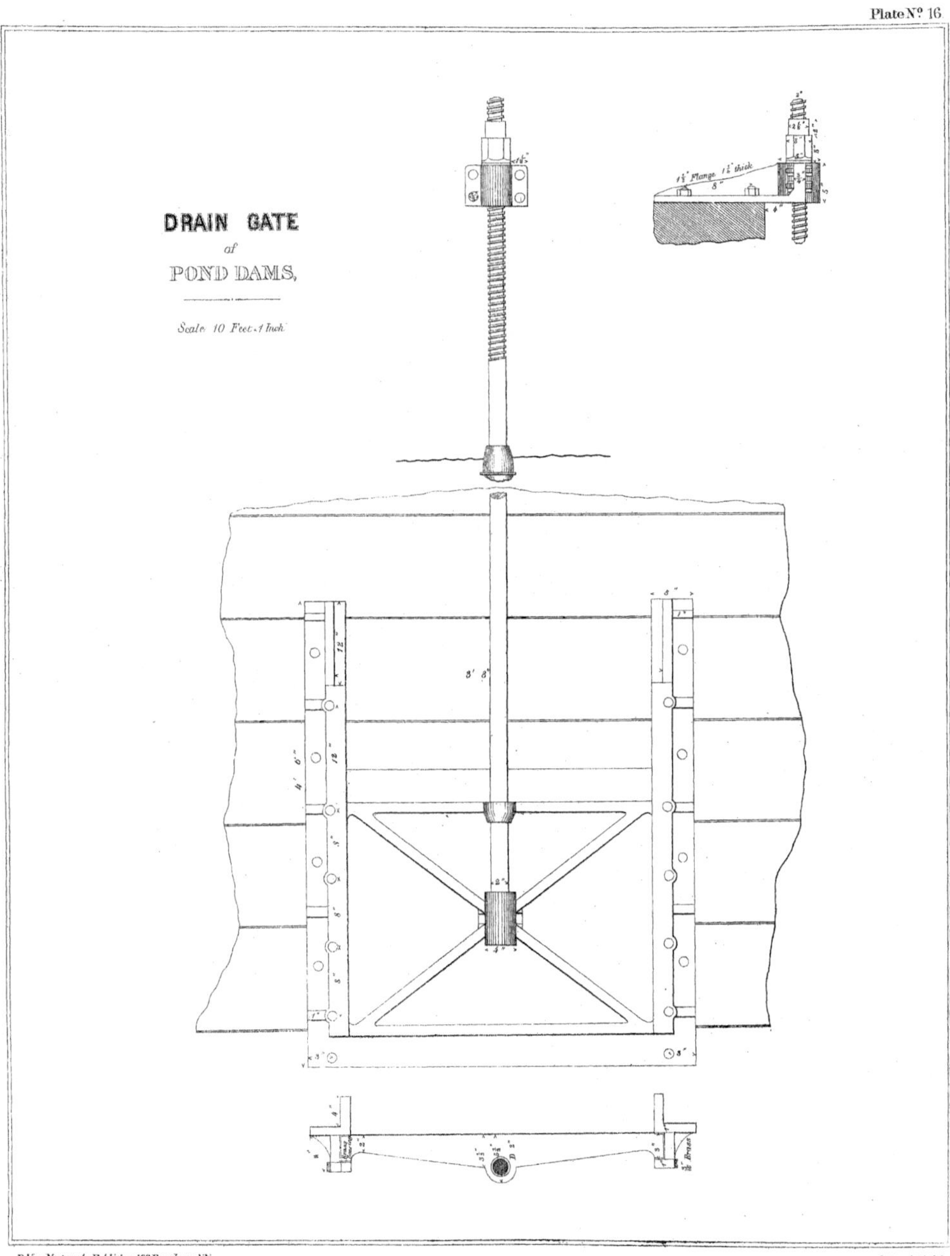


D. Van Nostrand, Publisher 192 Broadway N.Y.
J. Bien Lith. N.Y.

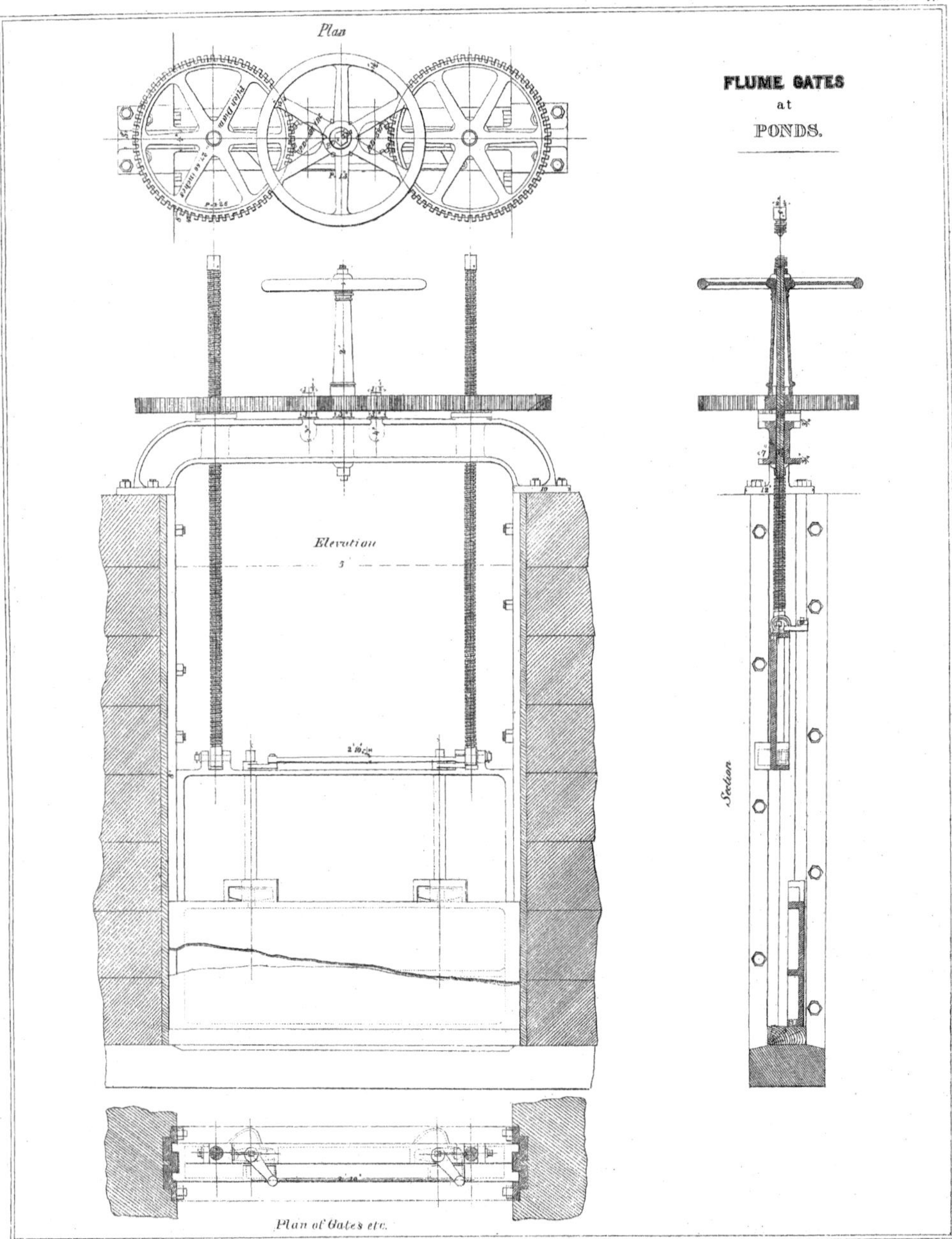


D. Van Nostrand, Publisher 192 Broadway, N.Y.
J. Bien Lith. N.Y.

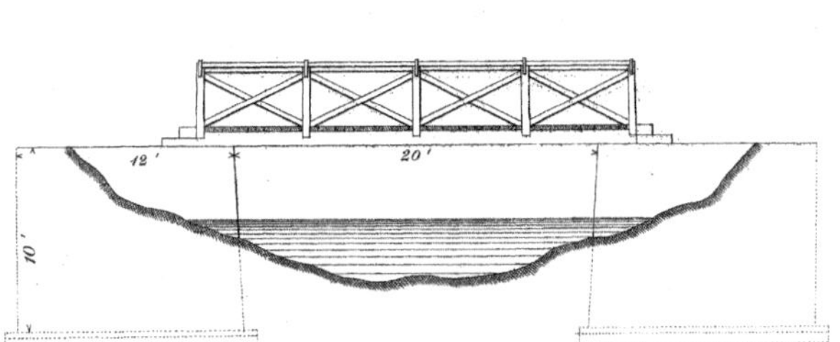


Bridge at head of Jamaica Pond.

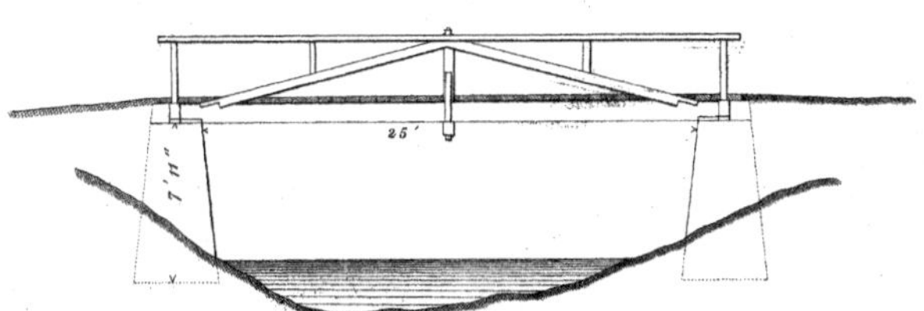


Bridge above Hempstead Pond.

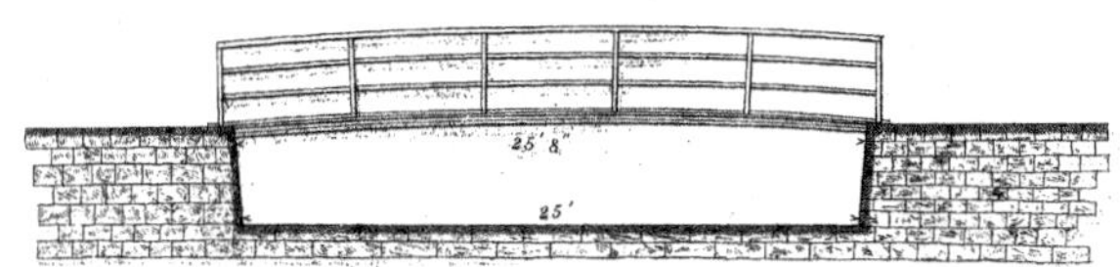


Bridge over Weir at Hempstead Pond.

D. Van Nostrand, Publisher 192 Broadway, N.Y.
J. Bien Lith. N.Y.

## PLAN AND ELEVATION
*of*
## GATE HOUSE AT PONDS.

*Scale 4 Ft. 1 Inch.*

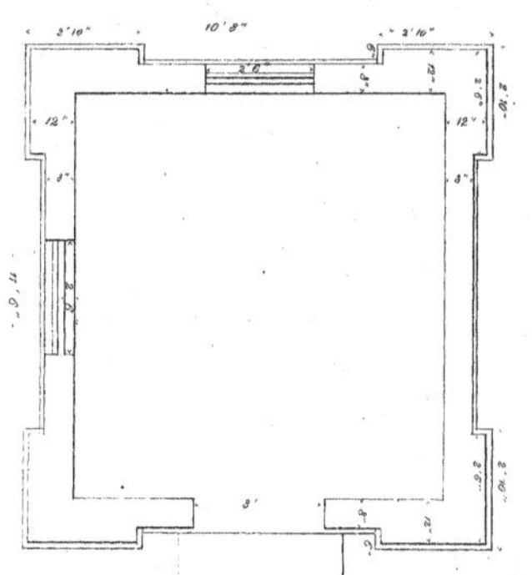

D. Van Nostrand, Publisher 192 Broadway, N.Y.

J. Bien Lith, N.Y.

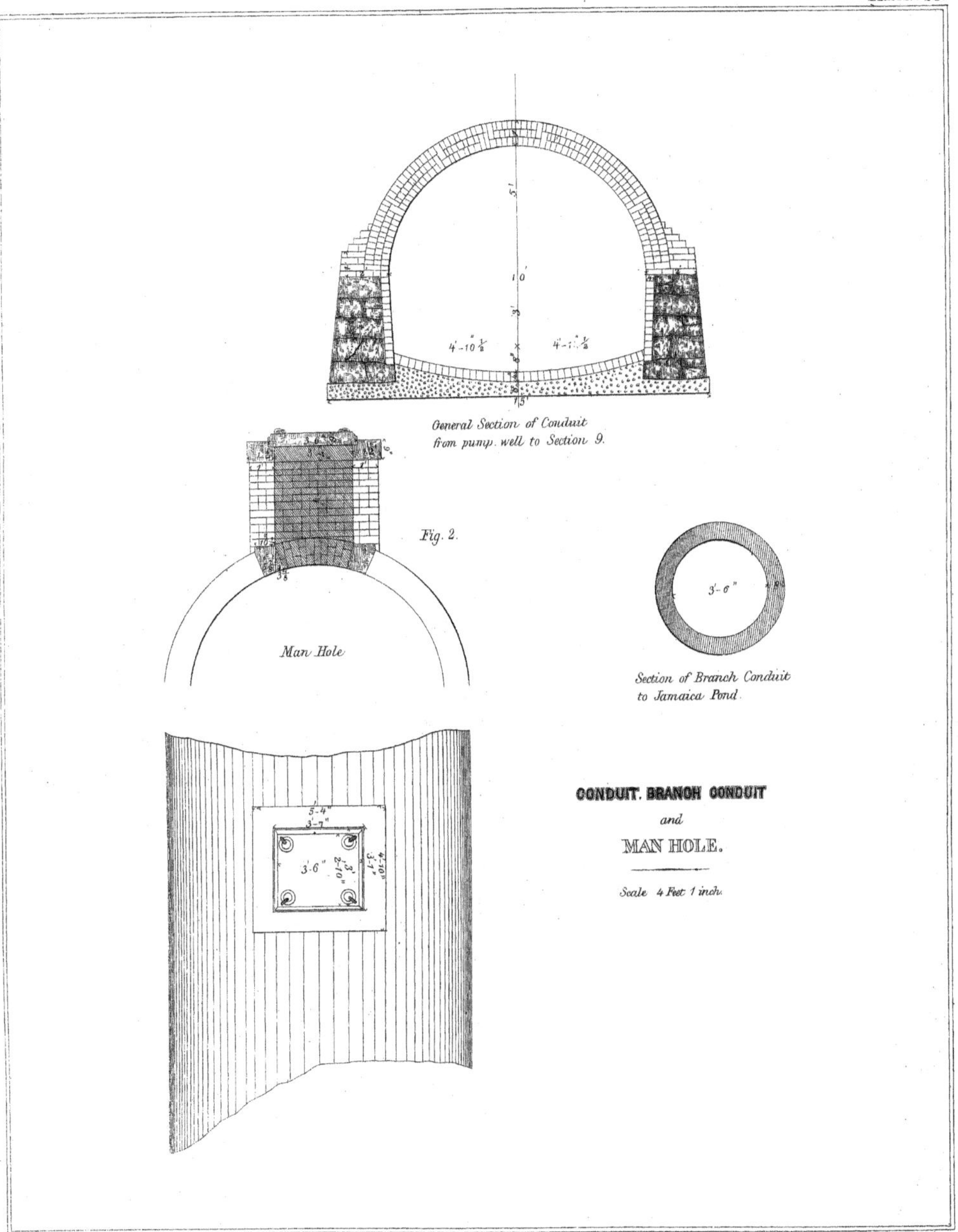

General Section of Conduit from pump. well to Section 9.

Section of Branch Conduit to Jamaica Pond.

CONDUIT, BRANCH CONDUIT and MAN HOLE.

Scale 4 Feet 1 inch.

D. Van Nostrand, Publisher 192 Broadway, N.Y.
J. Bien Lith. N.Y.

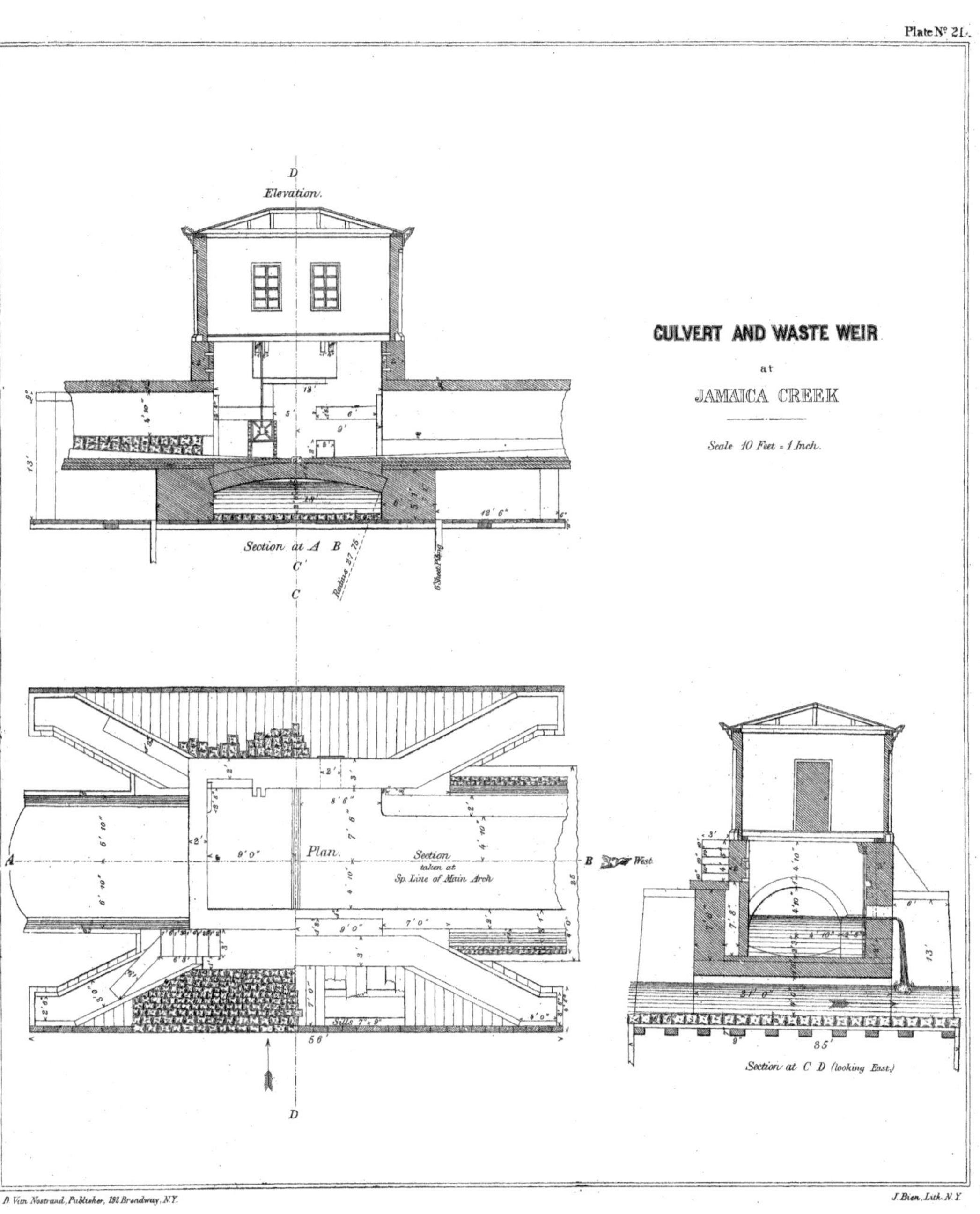
Plate Nº 21.
CULVERT AND WASTE WEIR
at
JAMAICA CREEK
Scale 10 Feet = 1 Inch.
Elevation.
Section at A B
Radius 27.75
Plan.
Section taken at Sp. Line of Main Arch
West
Section at C D (looking East.)
D. Van Nostrand, Publisher, 192 Broadway, N.Y.
J. Bien, Lith. N.Y.

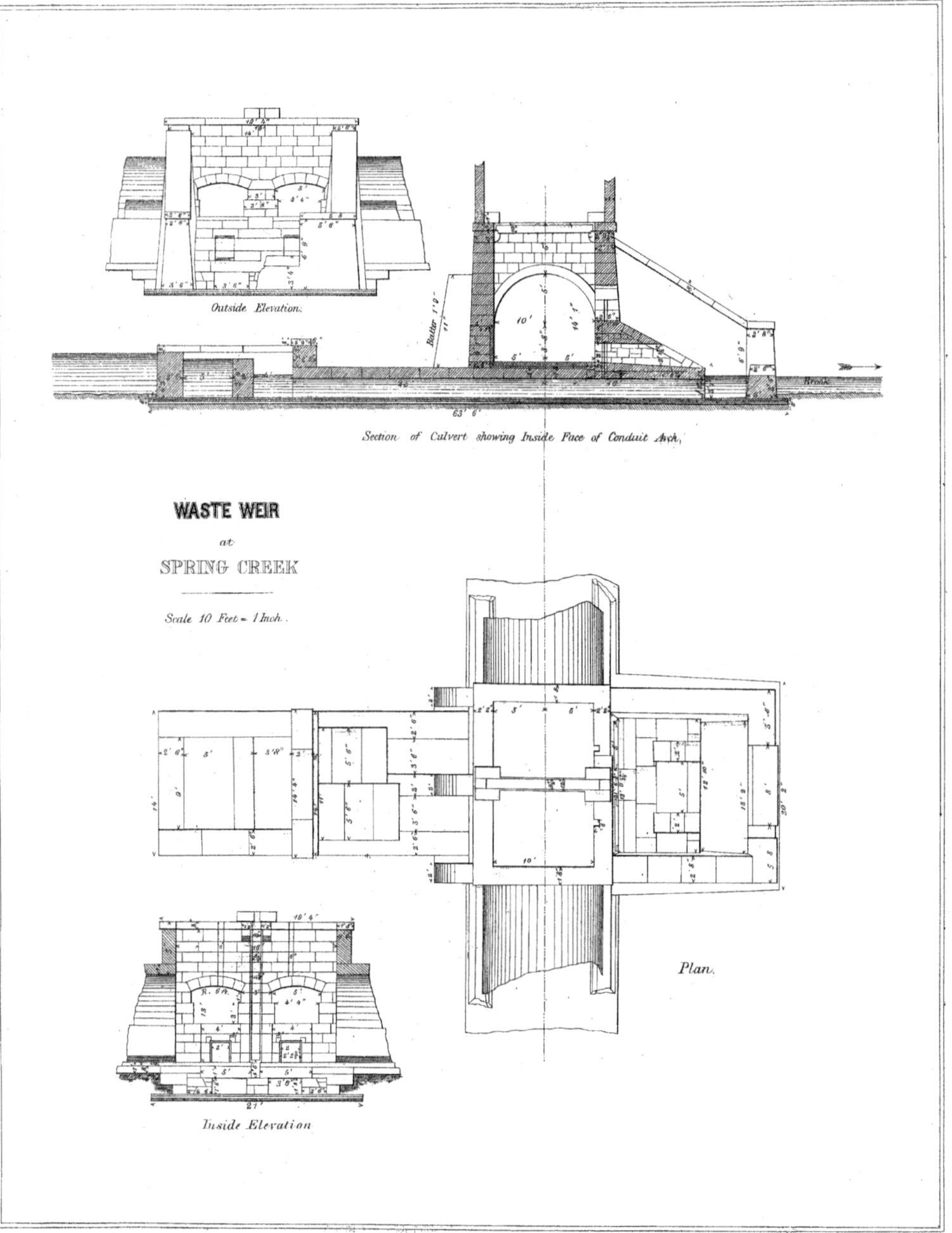


D. Van Nostrand, Publisher 192 Broadway, N.Y.

J. Bien Lith. N.Y.

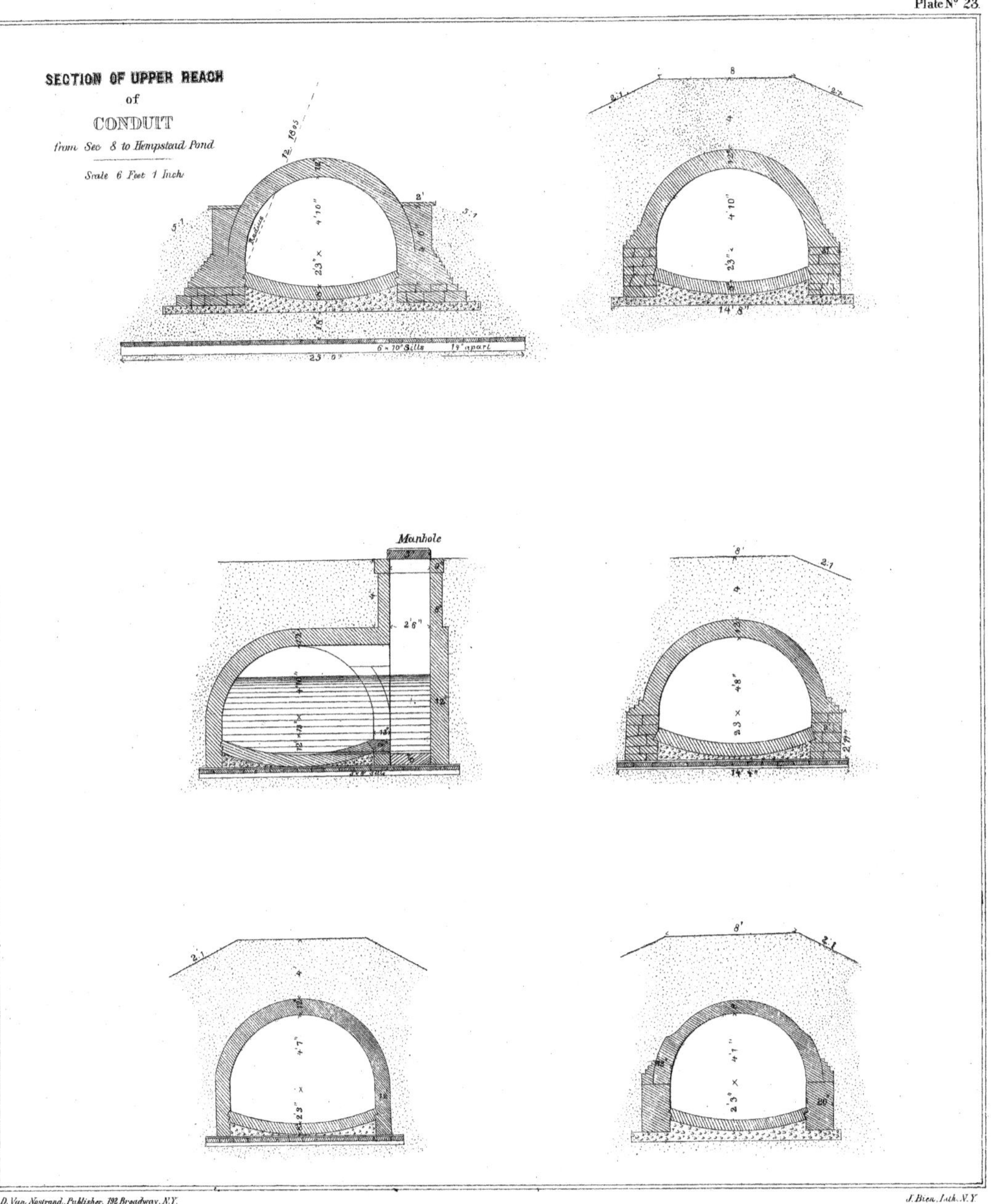


D. Van Nostrand, Publisher, 192 Broadway, N.Y.
J. Bien, Lith. N.Y.

*General South Elevation.*

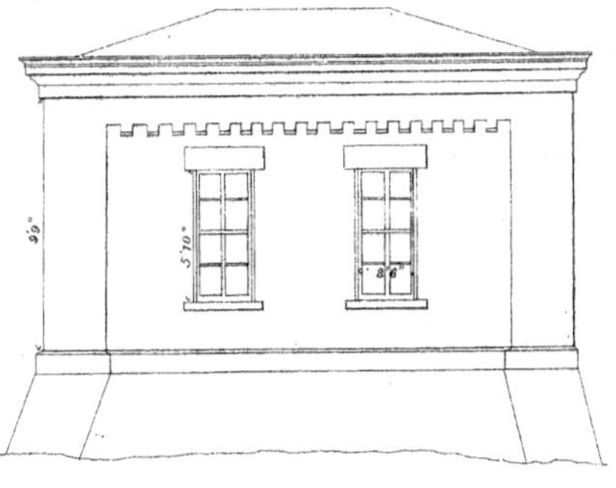

## Waste Weir House

*Scale 6 Feet 1 inch.*

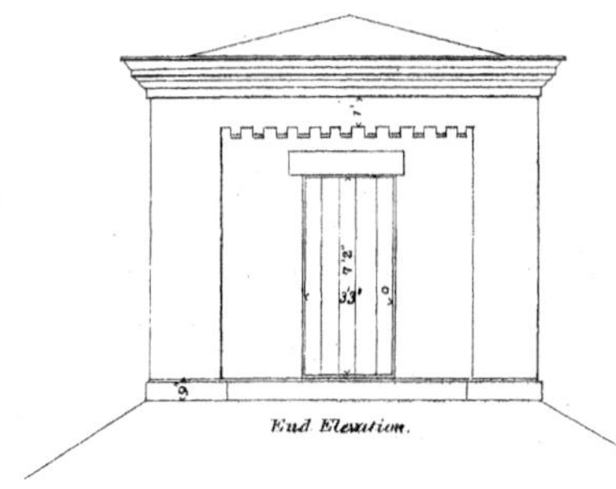


*End Elevation.*

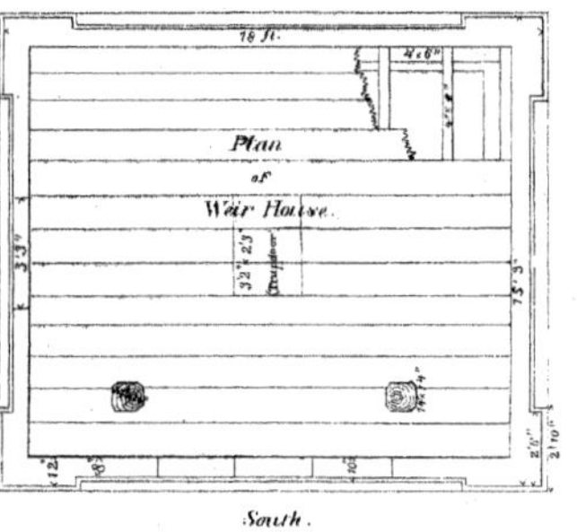


*South.*

D. Van Nostrand, Publisher 192 Broadway, N.Y.

J. Bien Lith. N.Y.

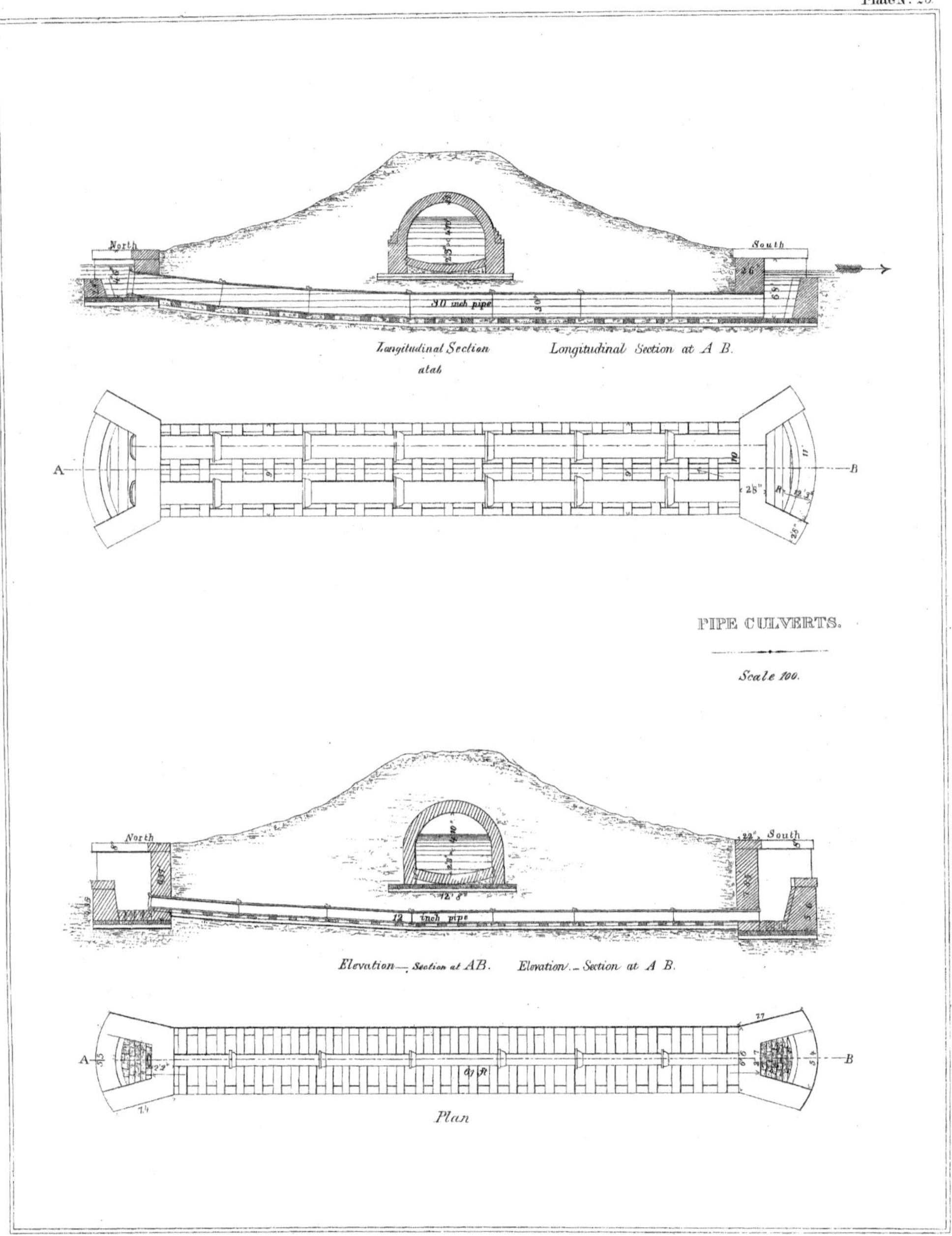


D. Van Nostrand, Publisher 192 Broadway, N.Y.
J. Bien Lith. N.Y.

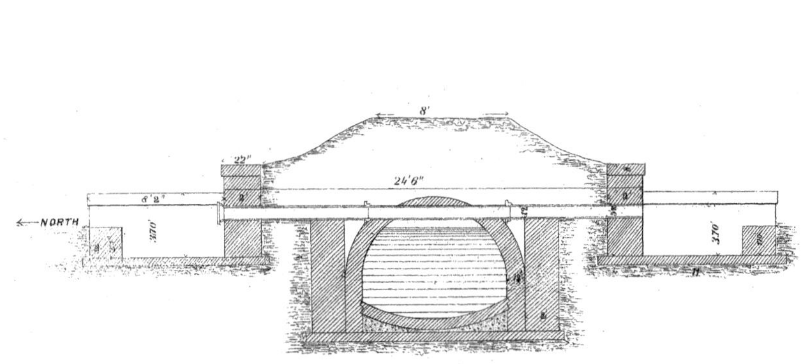


## PIPE CULVERT

Scale 6 Feet = 1 Inch.

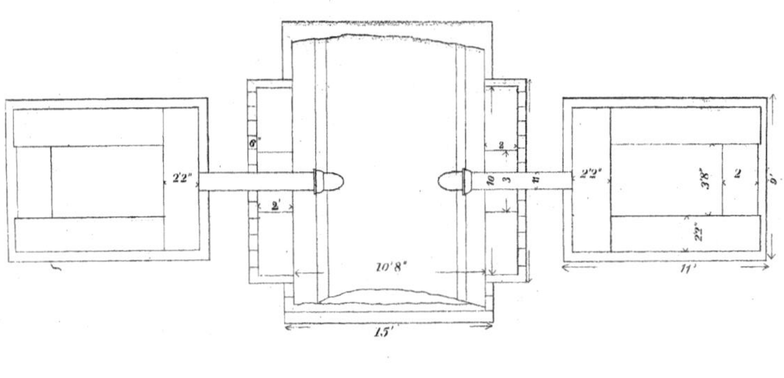


D. Van Nostrand, Publisher 192 Broadway, N.Y.

J. Bien Lith. N.Y.

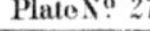

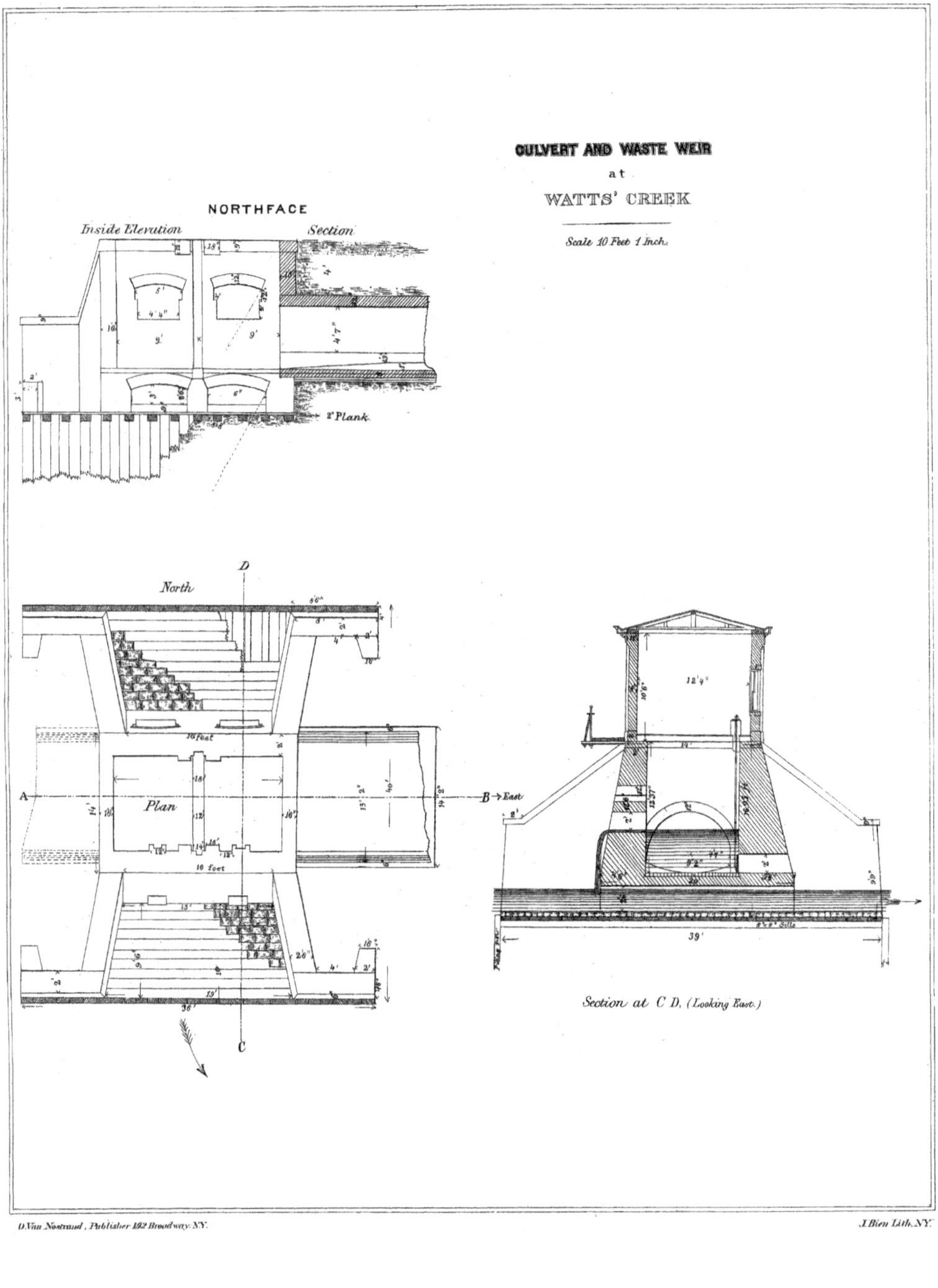

# CULVERT AND WASTE WEIR
at
# WATTS' CREEK

Scale 10 Feet 1 Inch.

NORTHFACE

*Inside Elevation* — *Section*

2" Plank.

North

Plan

A — B → East

D — C

Section at C D. (Looking East.)

D. Van Nostrand, Publisher 192 Broadway, N.Y. — J. Bien Lith. N.Y.

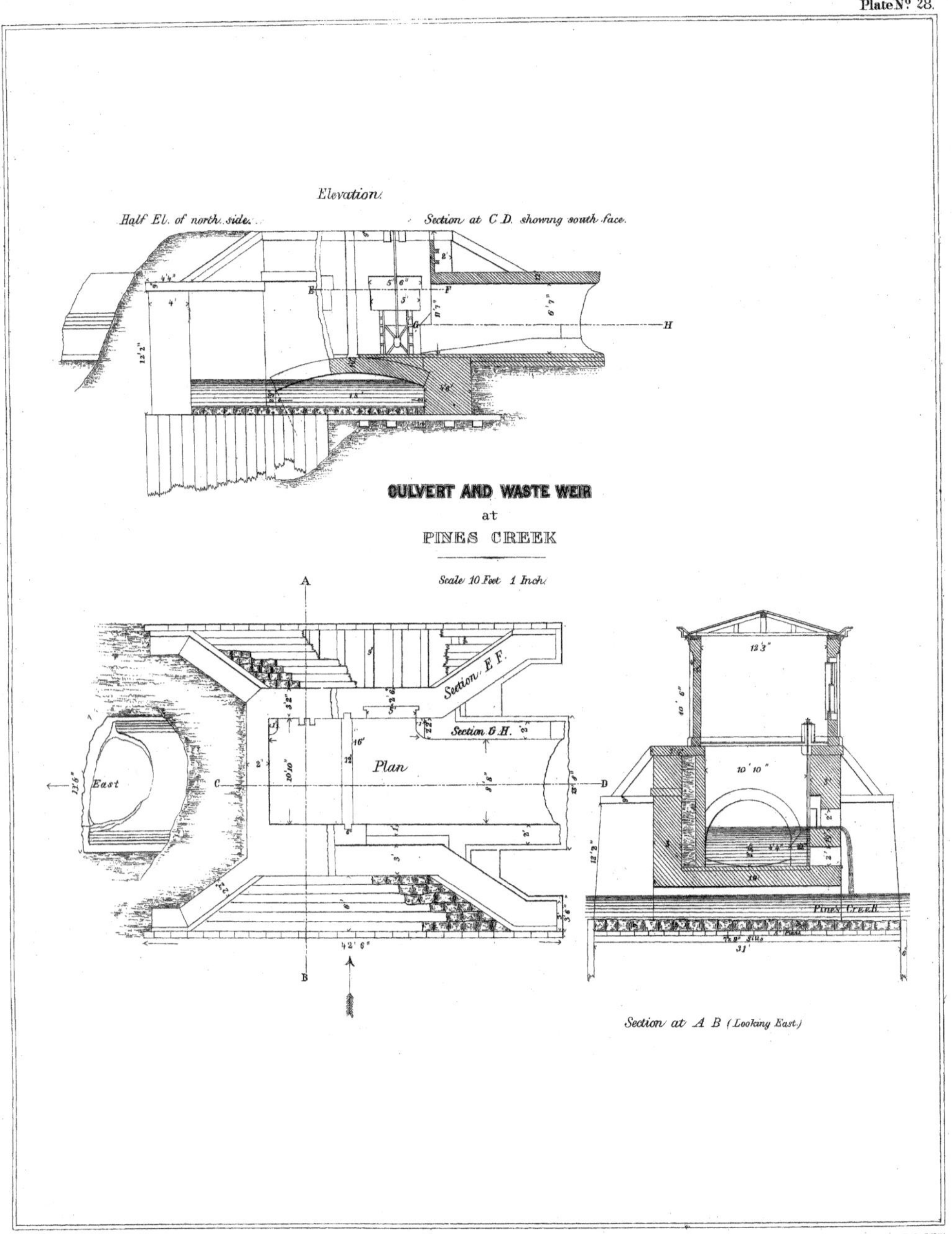


D. Van Nostrand, Publisher 192 Broadway, N.Y.

J. Bien Lith. N.Y.

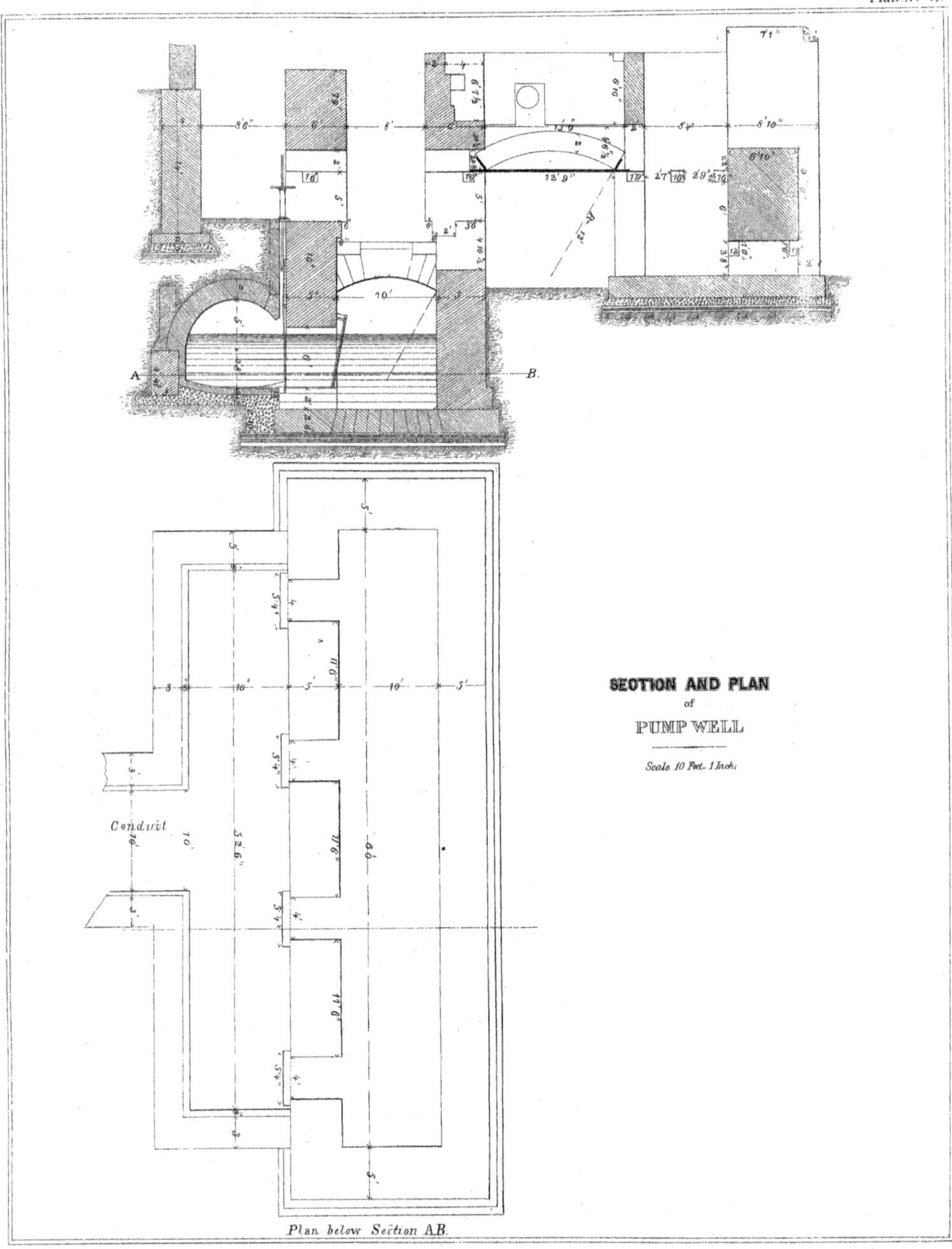


**SECTION AND PLAN**
of
**PUMP WELL**

Scale 10 Feet. 1 Inch.

D. Van Nostrand, Publisher 192 Broadway, N.Y.
J. Bien Lith. N.Y.

FOUNDATION
of
ENGINE Nº 2.
RIDGEWOOD

Scale 8 Feet 1 Inch.

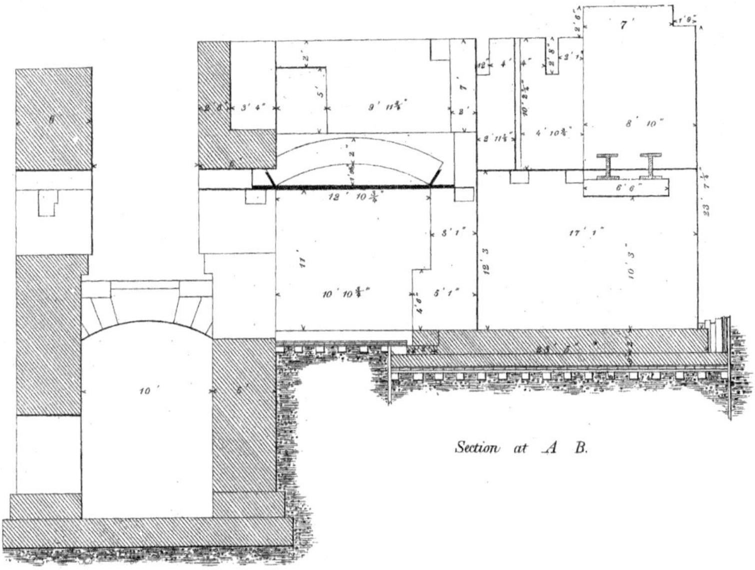

Section at A B.

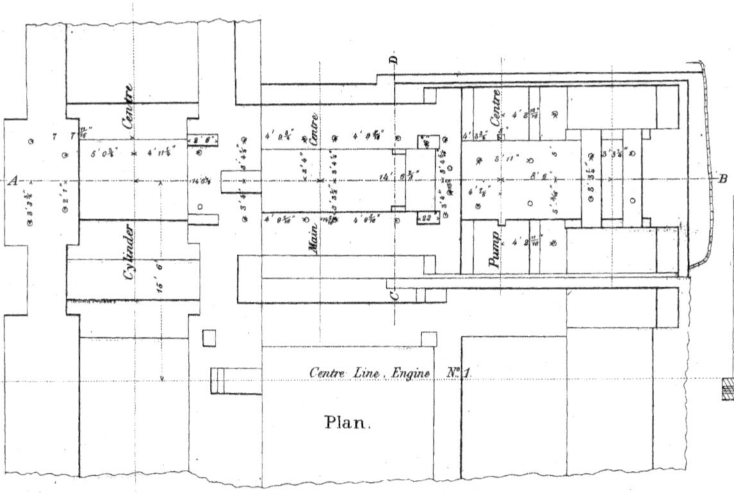

Plan.

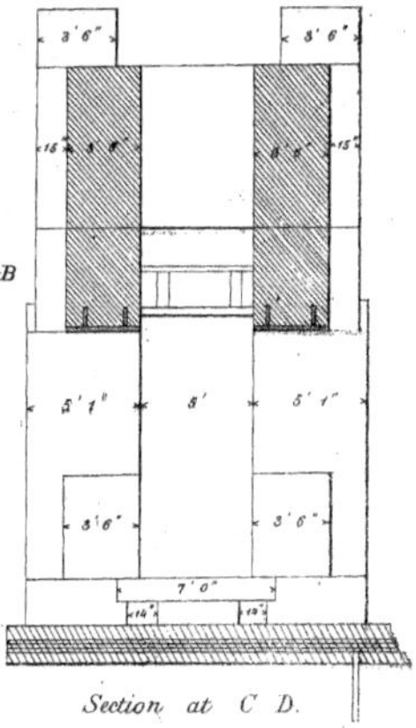

Centre Line, Engine Nº 1.

Section at C D.

D. Van Nostrand, Publisher, 192 Broadway, N.Y.

J. Bien, Lith. N.Y.

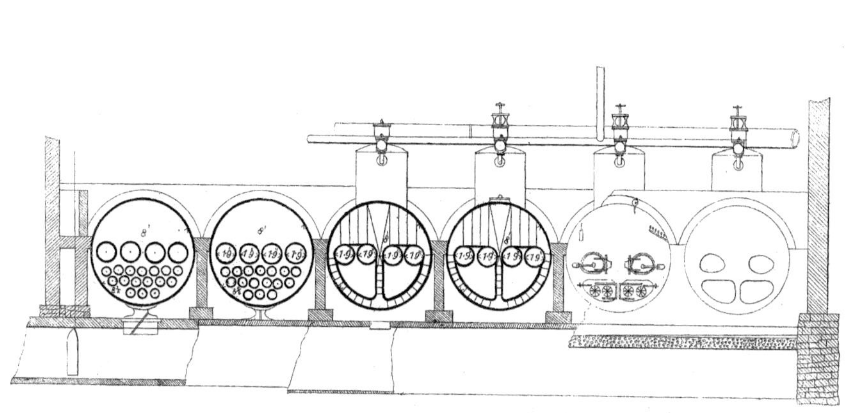

**BOILERS and FOUNDATIONS**

at

RIDGEWOOD ENGINE HOUSE

*Scale 8 Ft. 1 inch.*

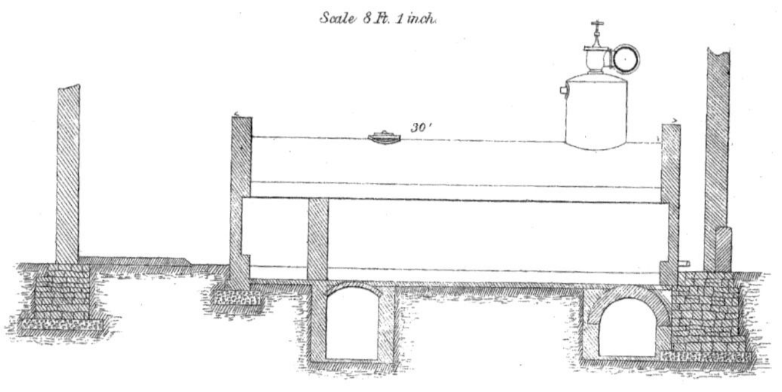


D. Van Nostrand, Publisher 192 Broadway N.Y.

J. Bien Lith. N.Y.

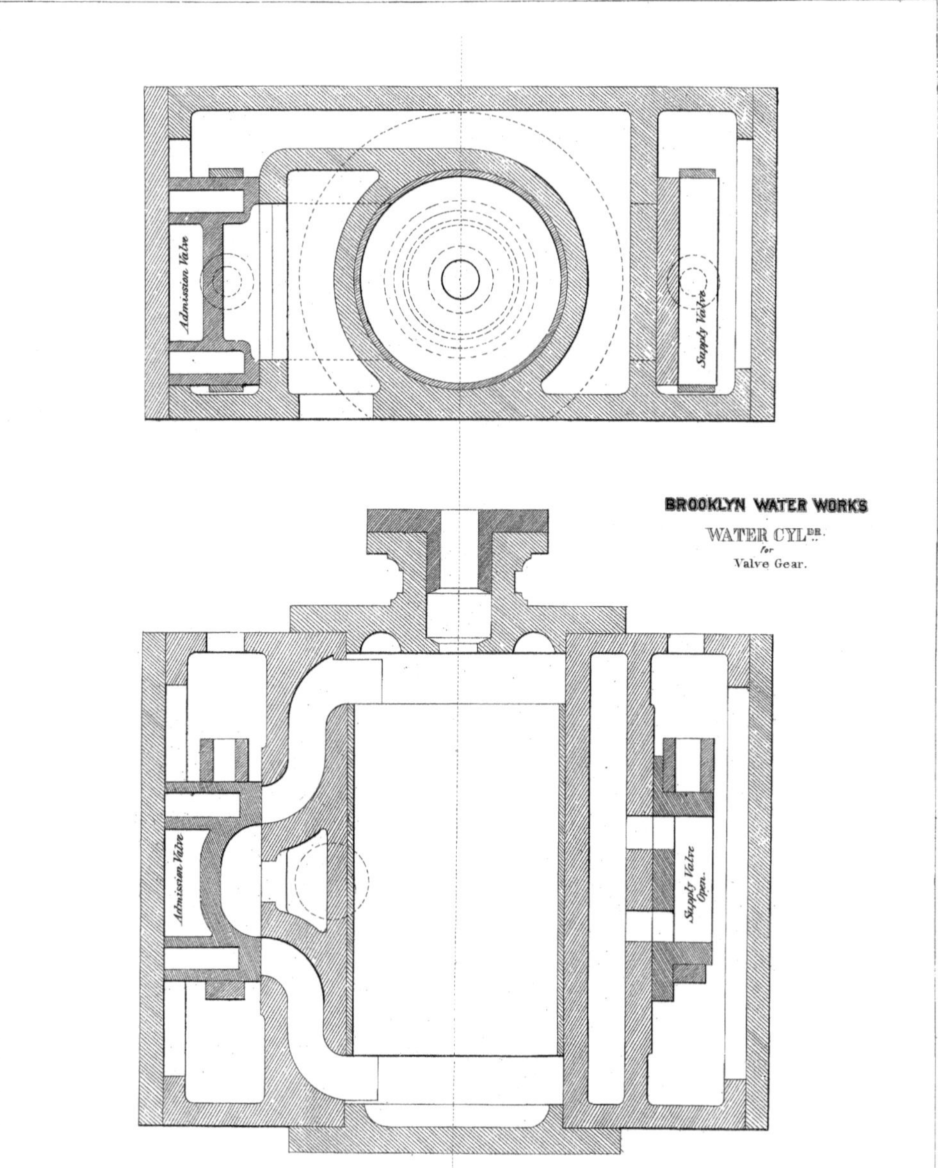


D. Van Nostrand, Publisher 192 Broadway. N.Y.

J. Bien Lith. N.Y.

Front Elevation. Front Elevation.

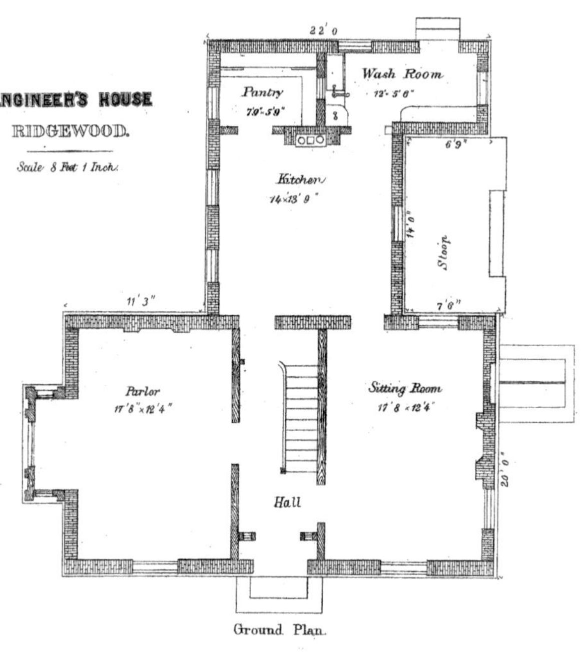


**ENGINEER'S HOUSE**

**RIDGEWOOD.**

*Scale 8 Feet 1 Inch.*

Ground Plan.

D. Van Nostrand, Publisher 192 Broadway, N.Y.

J. Bien Lith. N.Y.

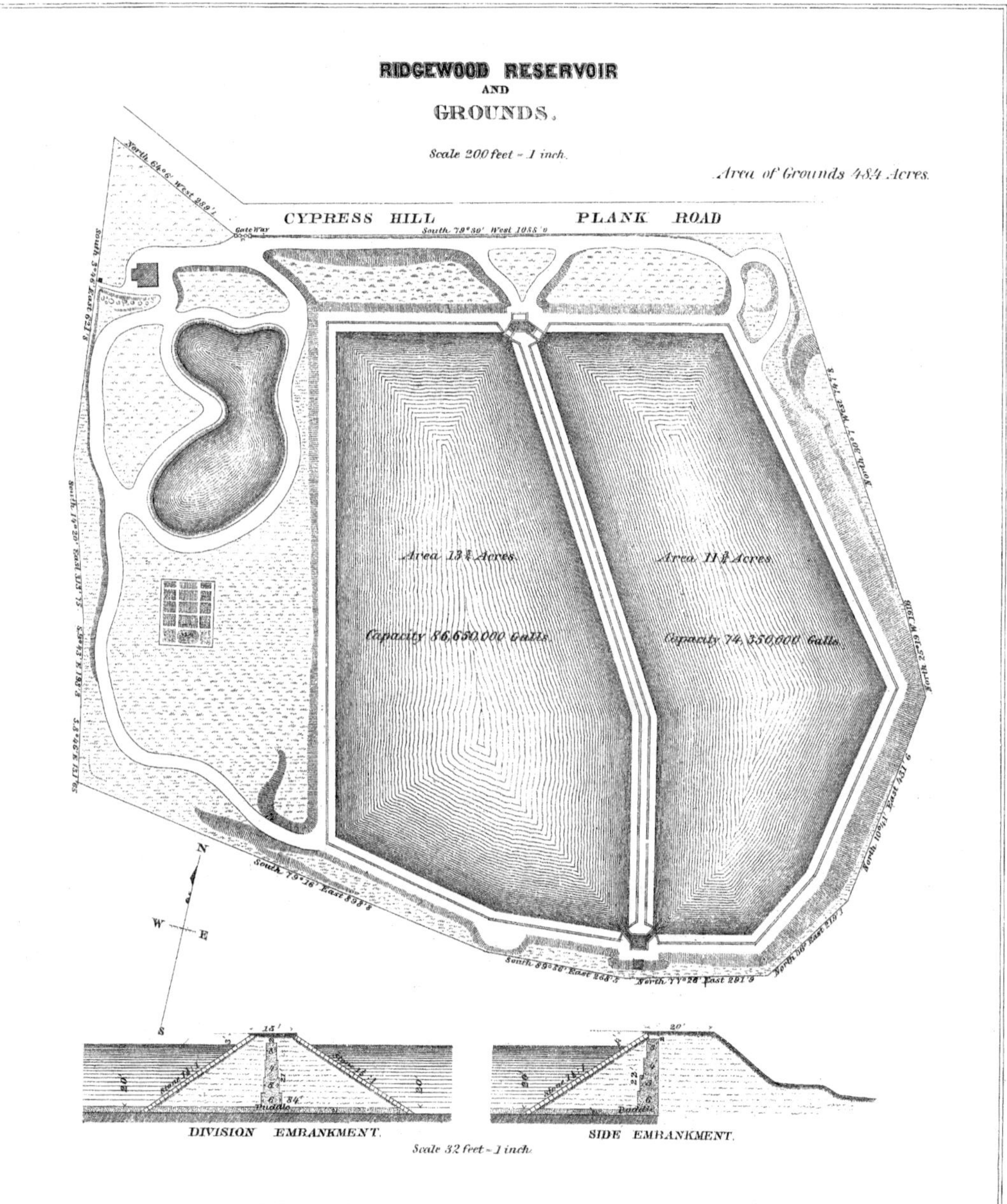


D. Van Nostrand, Publisher, 192 Broadway, N.Y.

J. Bien, Lith. N.Y.

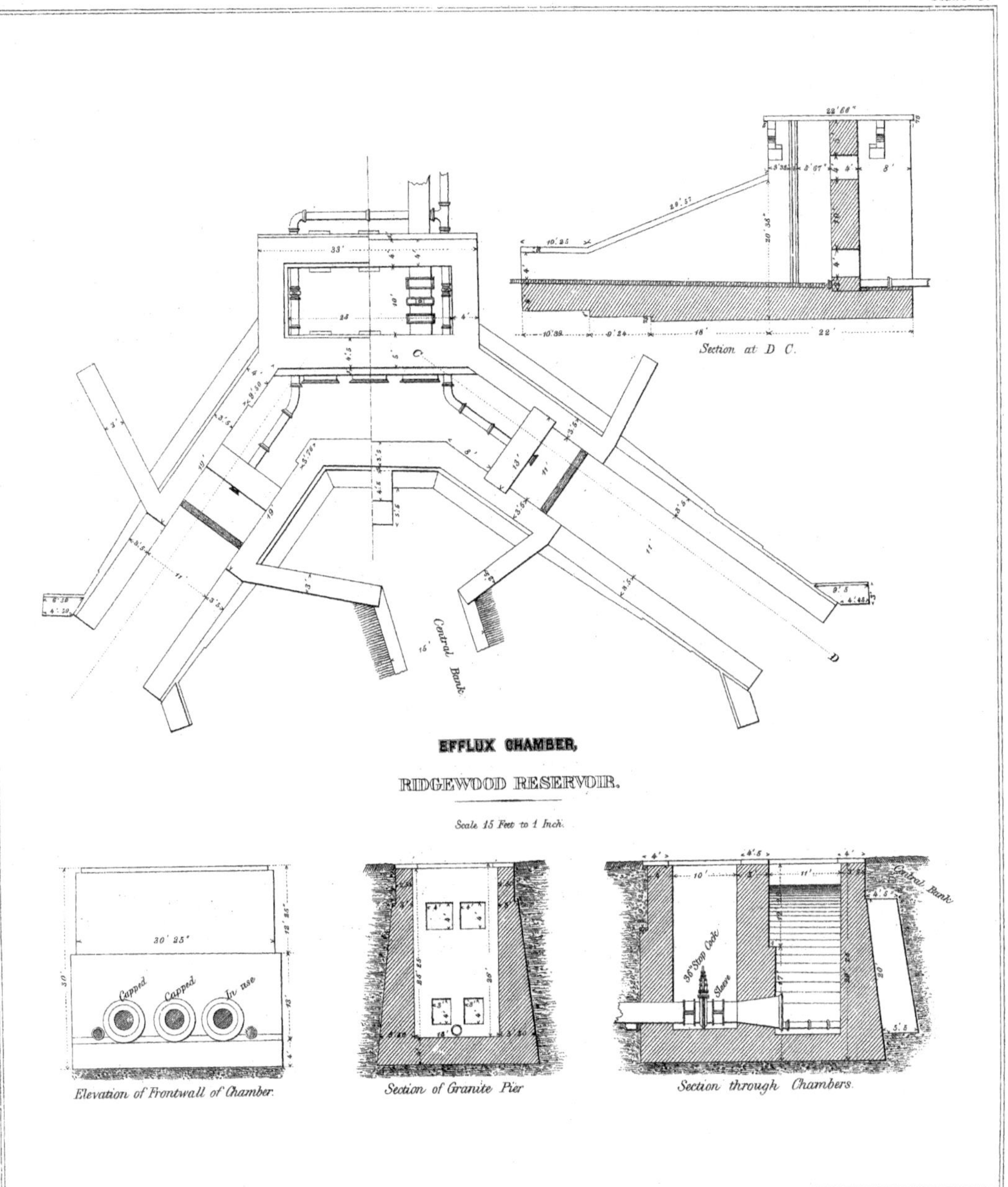


EFFLUX CHAMBER,

RIDGEWOOD RESERVOIR.

Scale 15 Feet to 1 Inch.

Elevation of Frontwall of Chamber.

Section of Granite Pier

Section through Chambers.

D. Van Nostrand, Publisher, 192 Broadway, N.Y.

J. Bien, Lith. N.Y.

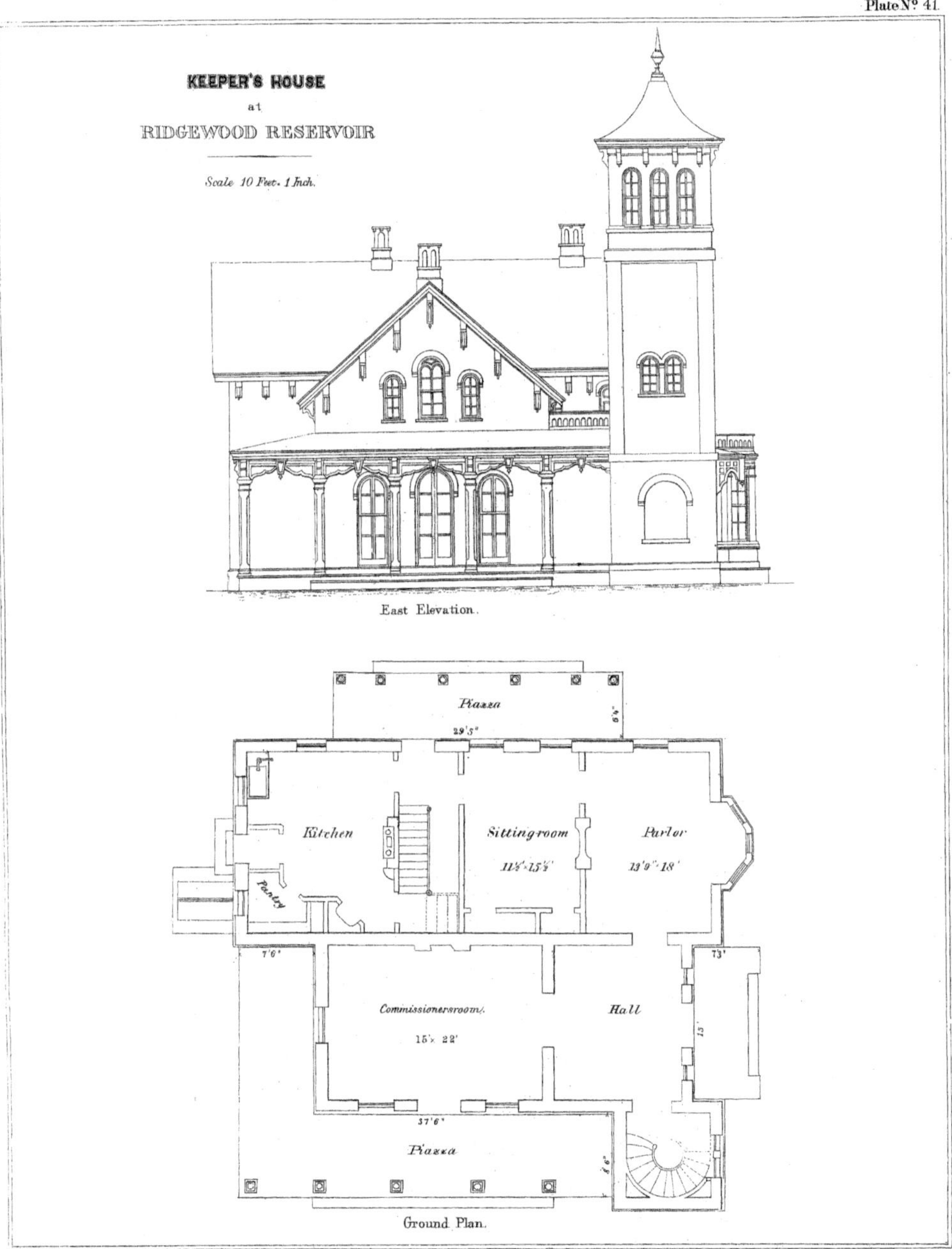


D. Van Nostrand, Publisher 192 Broadway N.Y.

J. Bien Lith. N.Y.

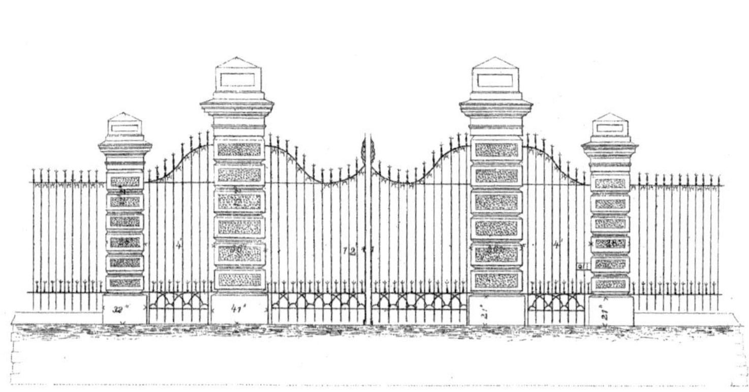

Entrance Gate at Ridgewood Reservoir.

Fences and Gates at Ponds.

**FENCES AND GATES**

Scale 6 Feet 1 Inch.

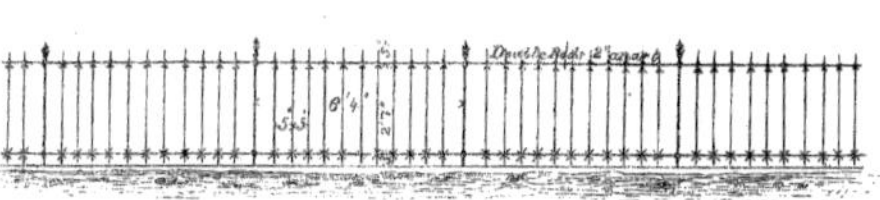

Fence on Coping around Ridgewood Reservoir.

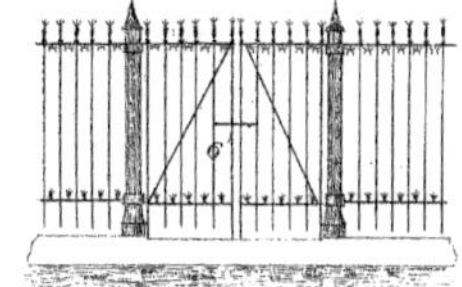

Gate at Mt. Prospect Reservoir.

D. Van Nostrand, Publisher 192 Broadway, N.Y.

J. Bien Lith. N.Y.

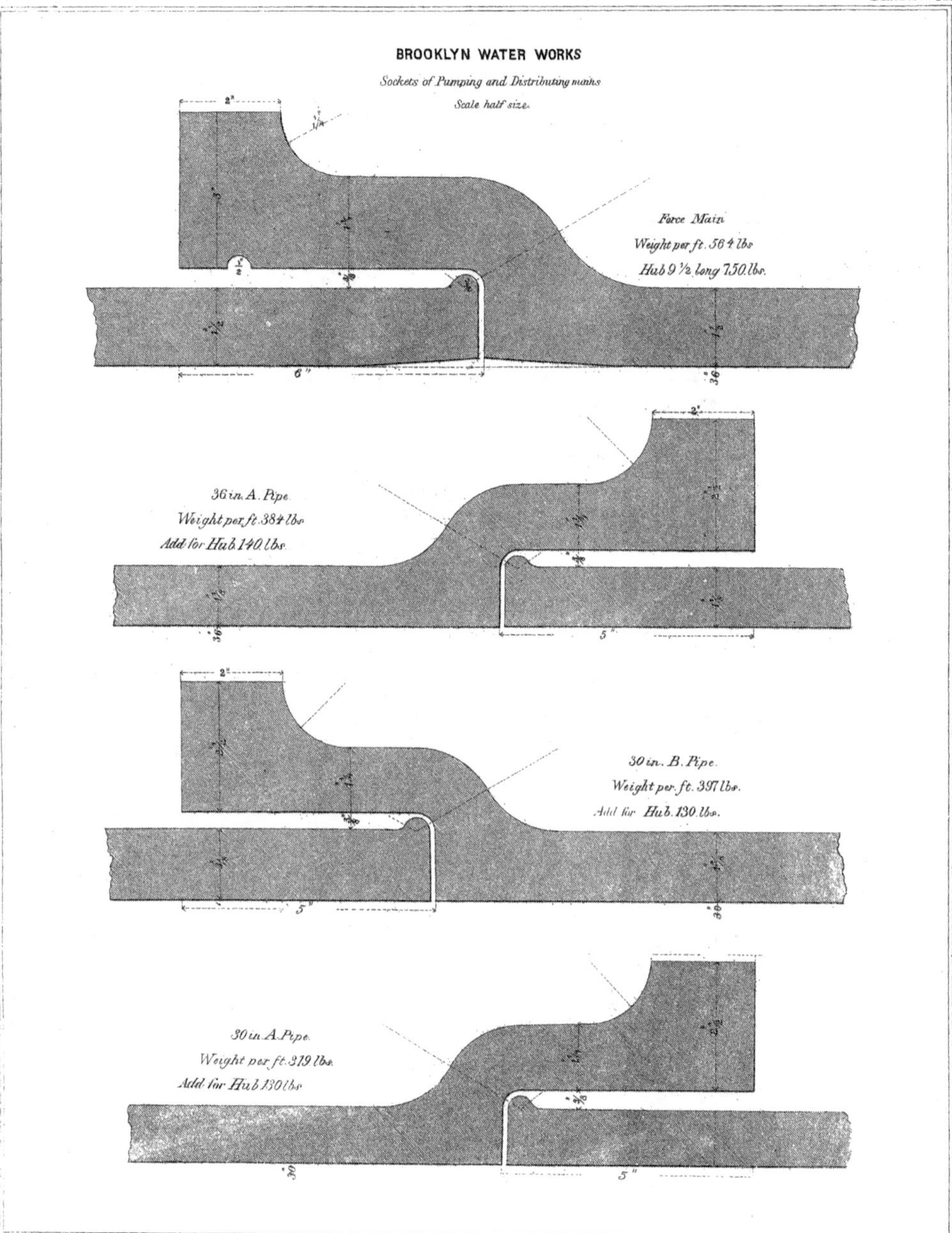


D. Van Nostrand, Publisher 192 Broadway, N.Y.

J. Bien Lith. N.Y.

**BROOKLYN WATER WORKS.**

*Sockets of Distribution pipes.*

*Scale half size.*

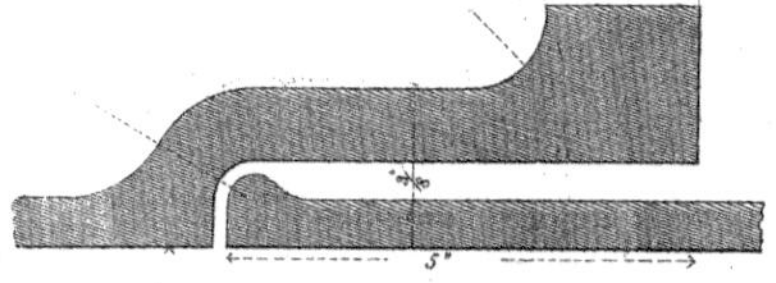


8 in. "A" pipe - per foot, 44 lbs
add to full length for hub, 30 lbs.

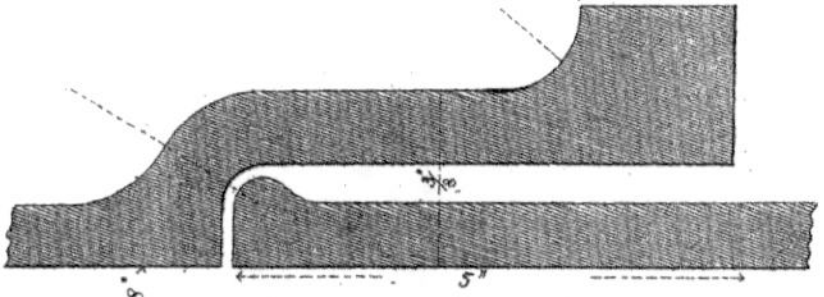


8 in. "B" pipe - per foot, 52 lbs
add to full length for hub, 30 lbs.

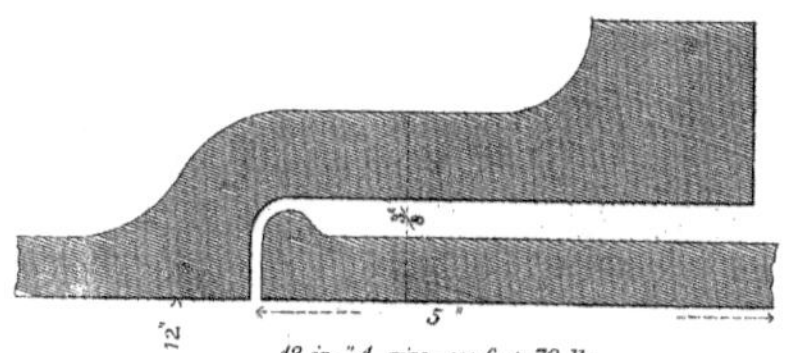


12 in. "A" pipe - per foot, 70 lbs
add to full length for hub, 50 lbs.

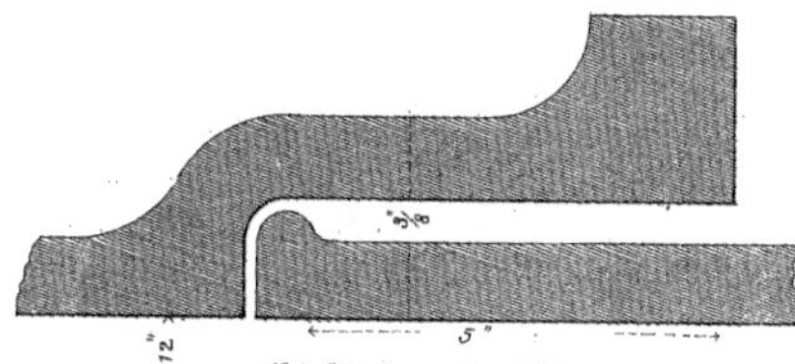


12 in. "B" pipe - per foot, 86 lbs
add to full length for hub, 50 lbs.

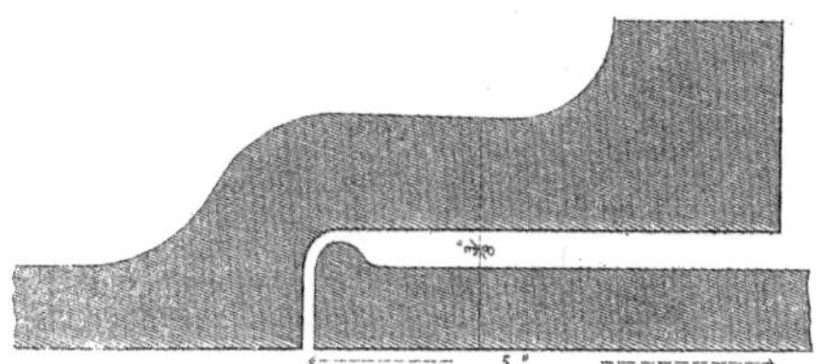


20 in. "A" pipe - per foot, 167 lbs.
add to full length for hub, 96 lbs.

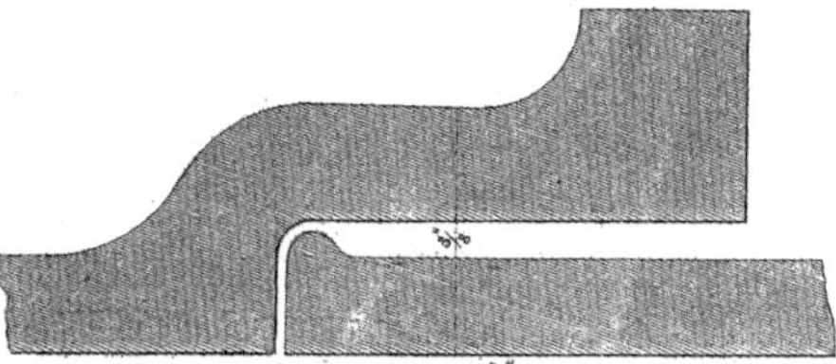


20 in. "B" pipe - per foot, 200 lbs.
add to full length for hub, 96 lbs.

D. Van Nostrand, Publisher 192 Broadway, N.Y.

J. Bien Lith. N.Y.

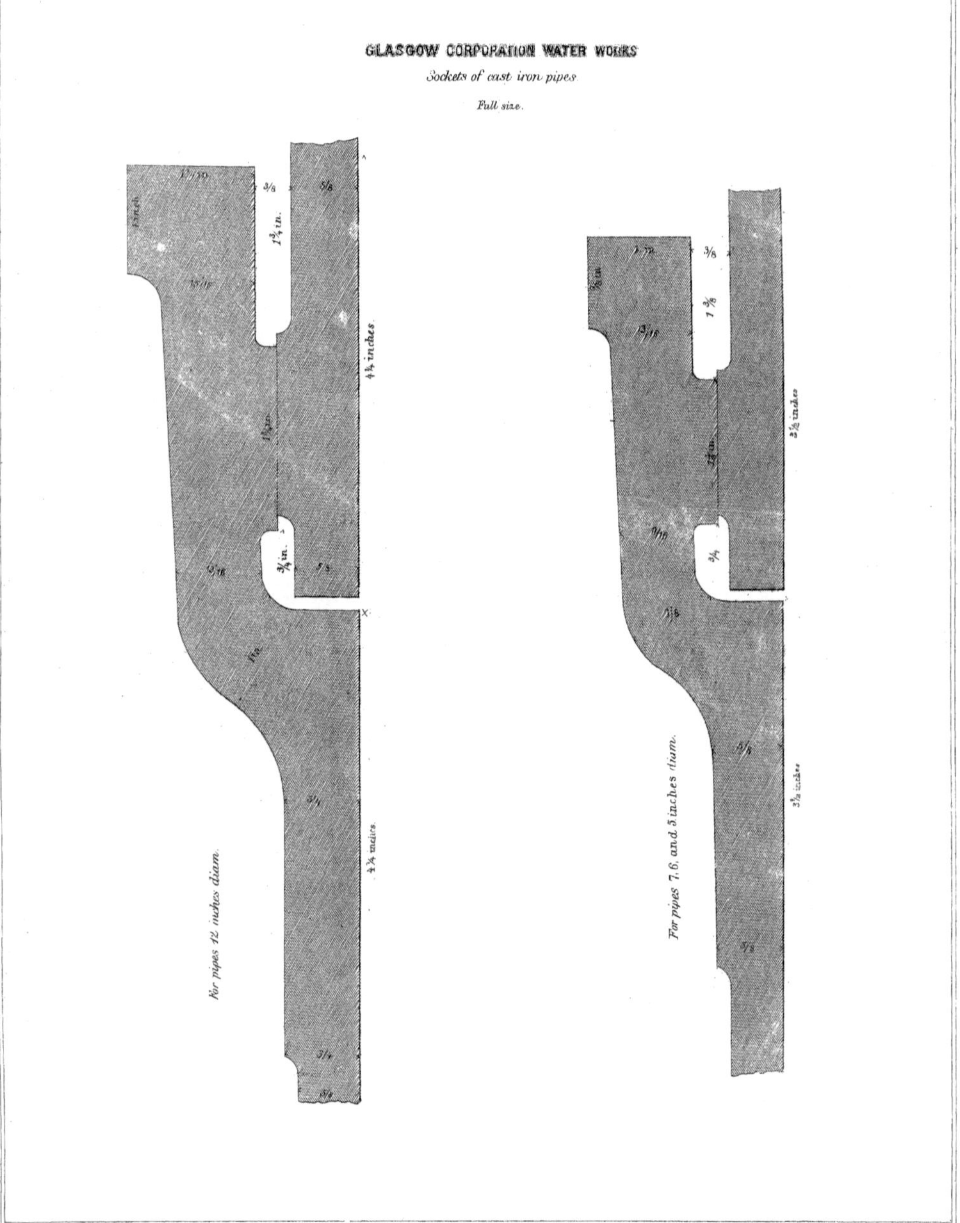


D. Van Nostrand, Publisher 192 Broadway, N.Y.

J. Bien Lith. N.Y.

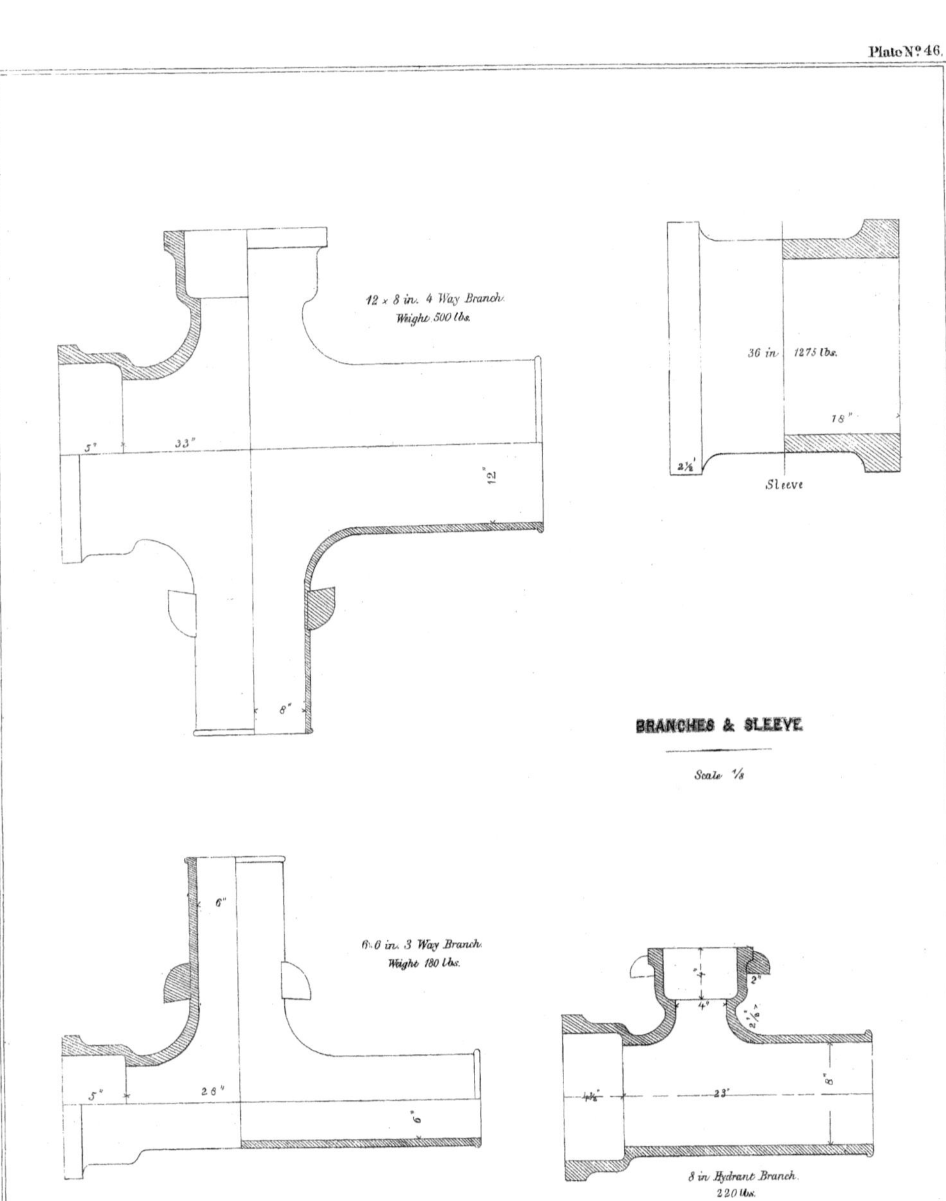


D. Van Nostrand, Publisher 192 Broadway N.Y. J. Bien Lith. N.Y.

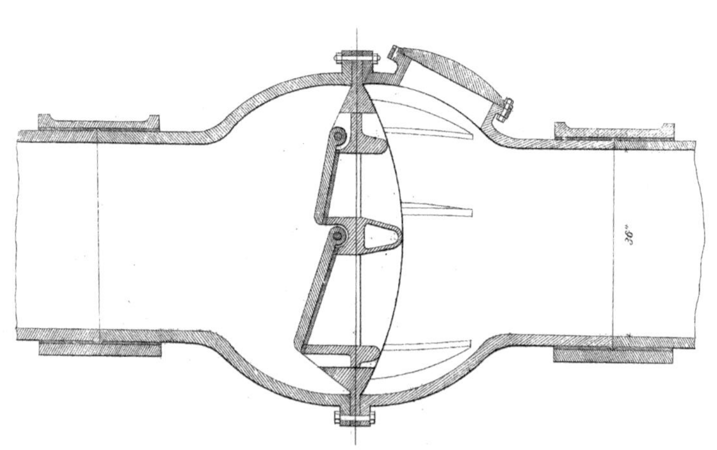


**CHECK VALVE**
for
FORCE MAINS

*Scale 20 Feet. 1 Inch.*

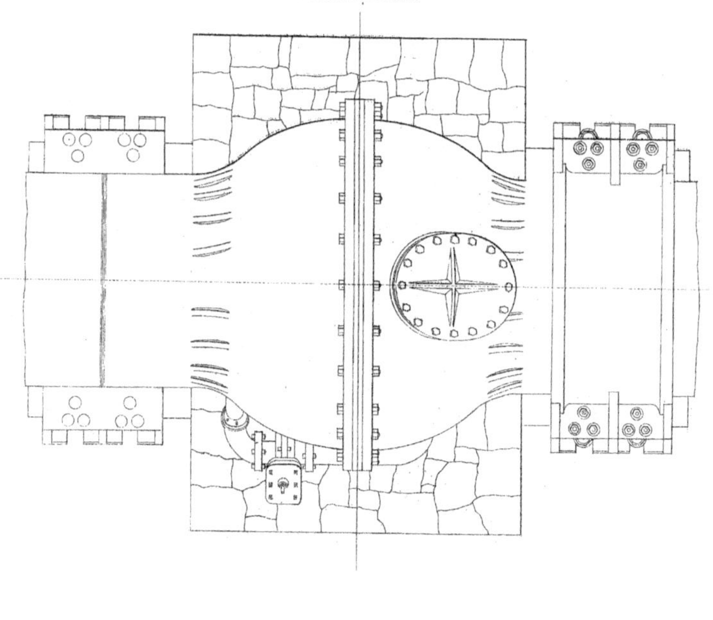

D. Van Nostrand, Publisher 192 Broadway. N.Y.

J. Bien Lith. N.Y.

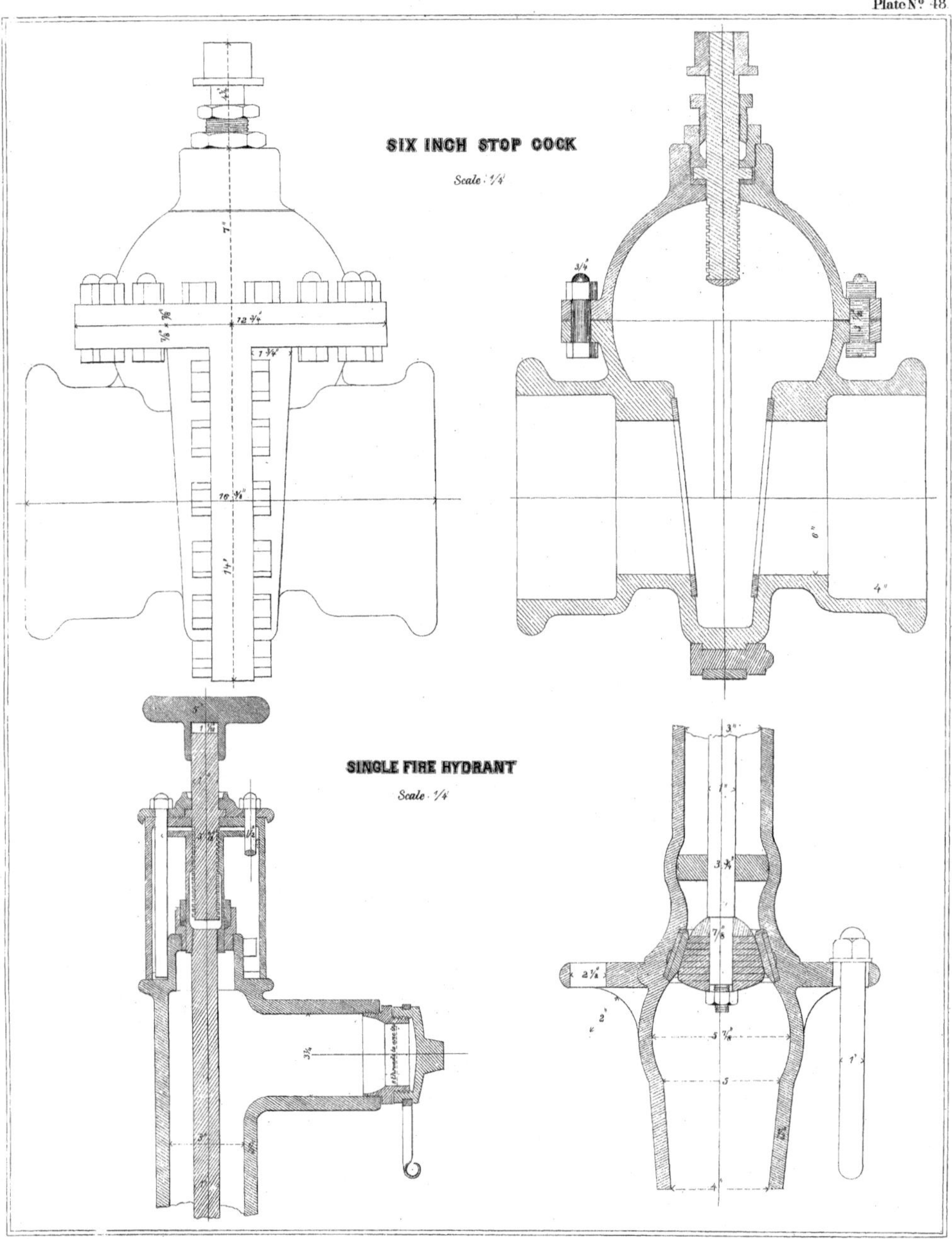


D. Van Nostrand, Publisher 192 Broadway, N.Y.
J. Bien Lith. N.Y.

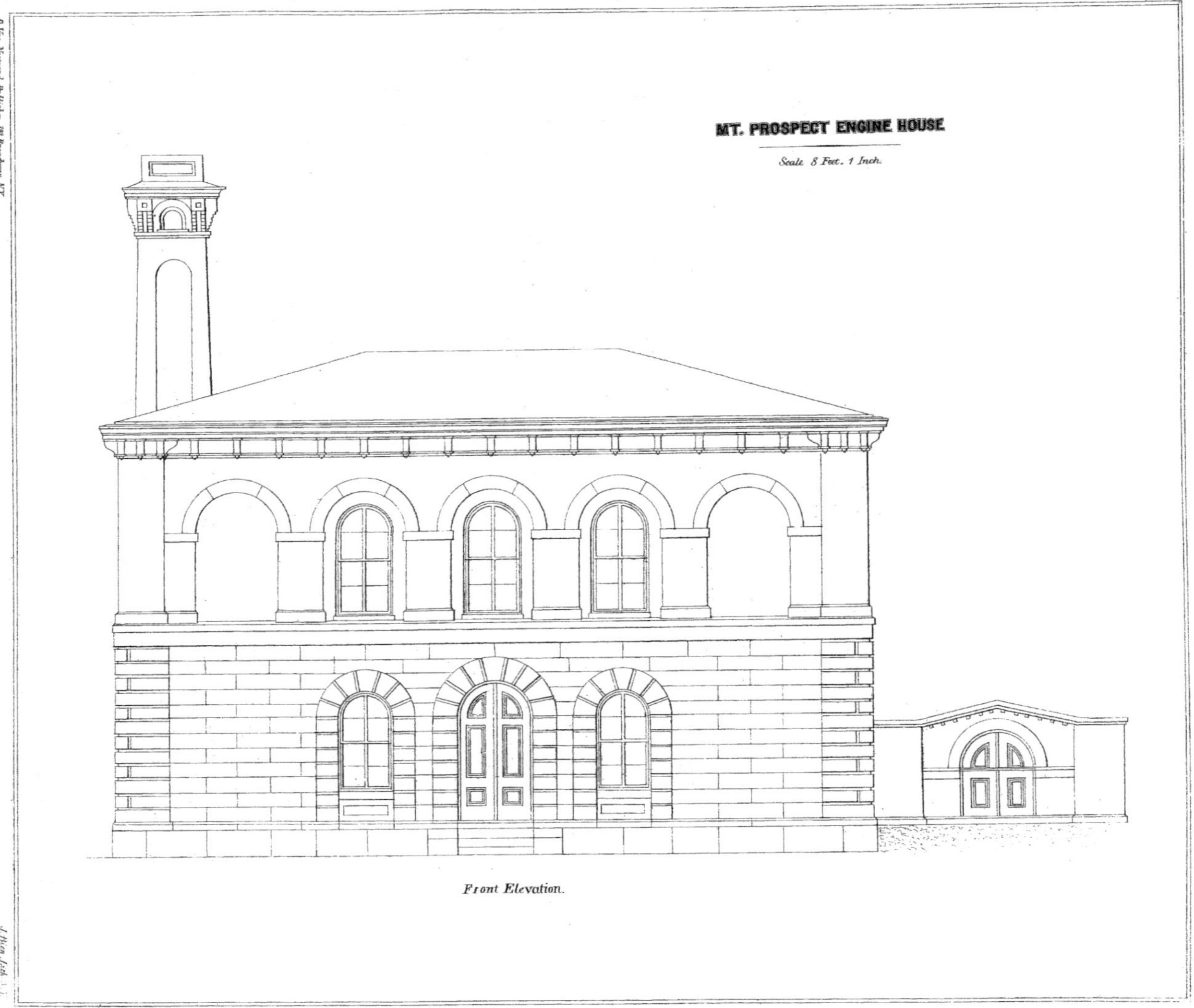


D. Van Nostrand, Publisher, 192 Broadway, N.Y.
J. Bien, Lith. N.Y.

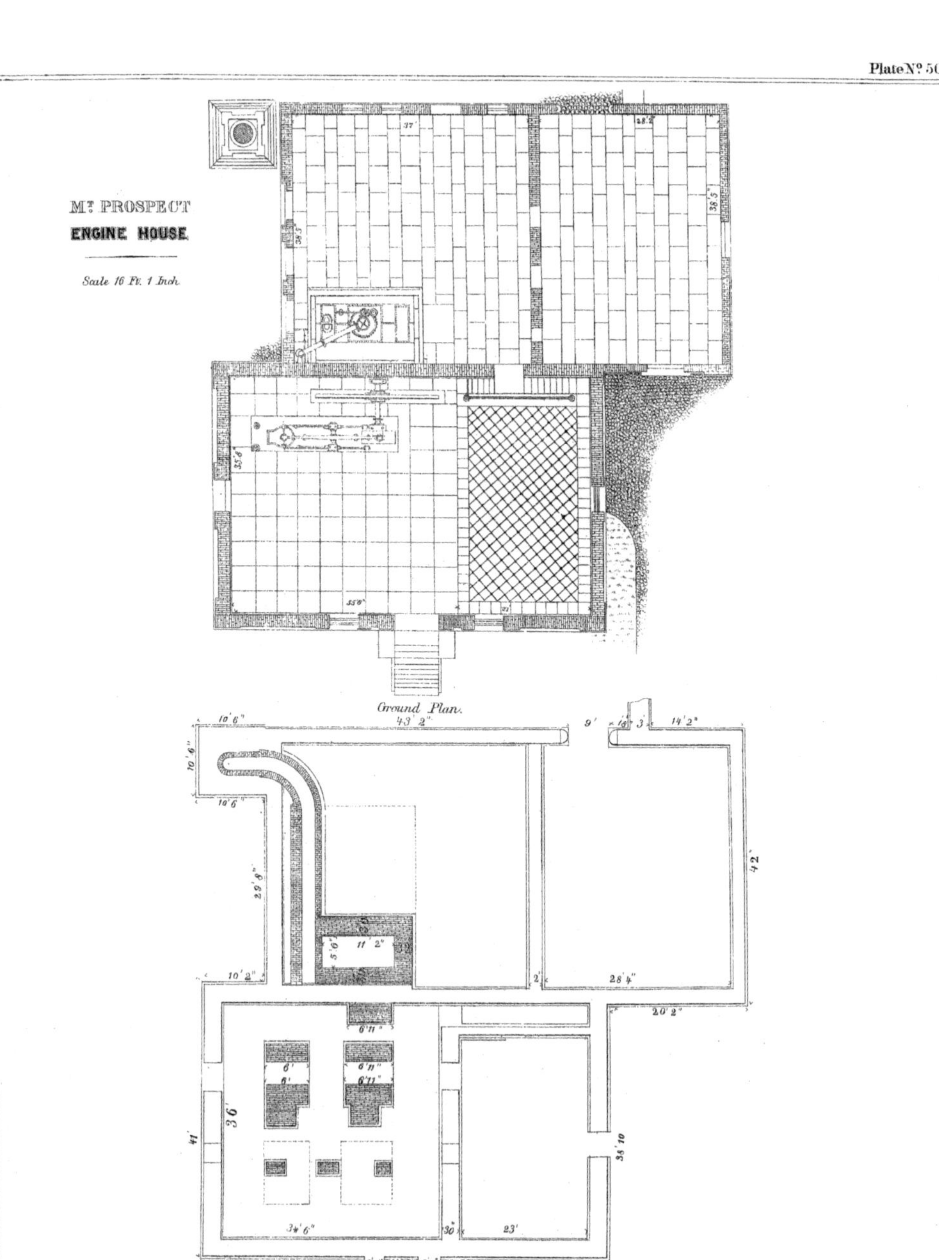


D. Van Nostrand, Publisher 192 Broadway, N.Y.

J. Bien Lith. N.Y.

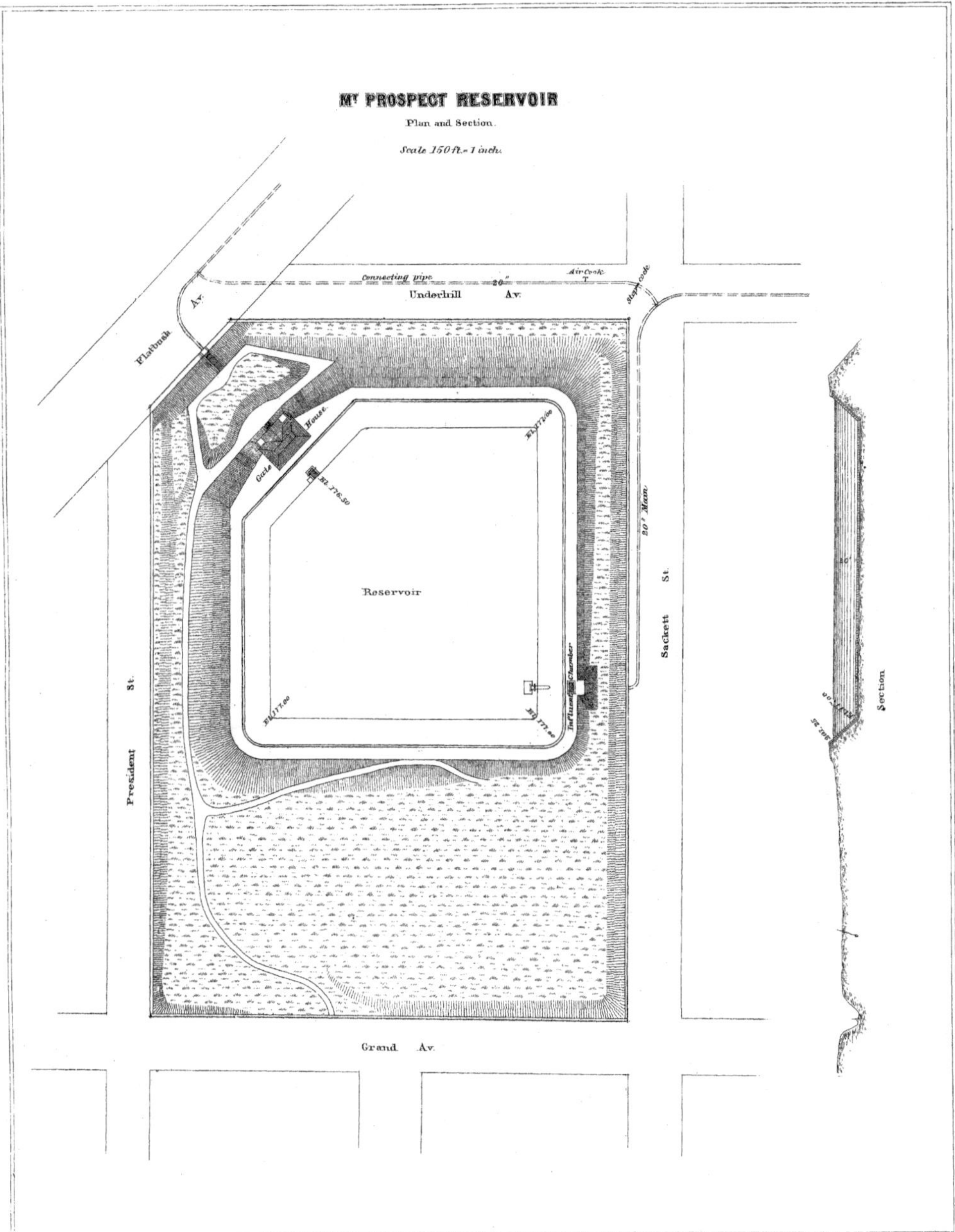


D. Van Nostrand, Publisher 192 Broadway, N.Y.
J. Bien Lith. N.Y.

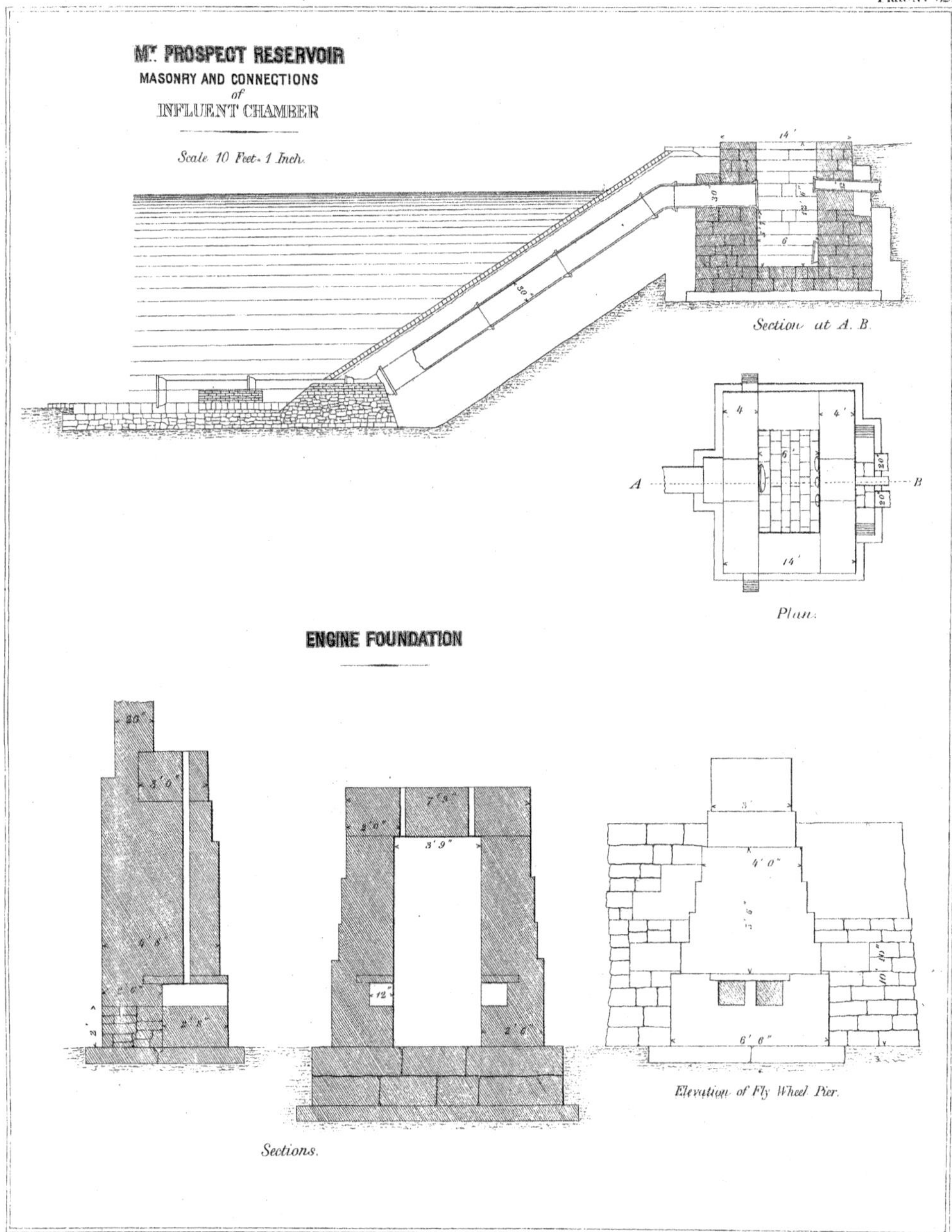


D. Van Nostrand, Publisher 192 Broadway, N.Y.

J. Bien Lith. N.Y.

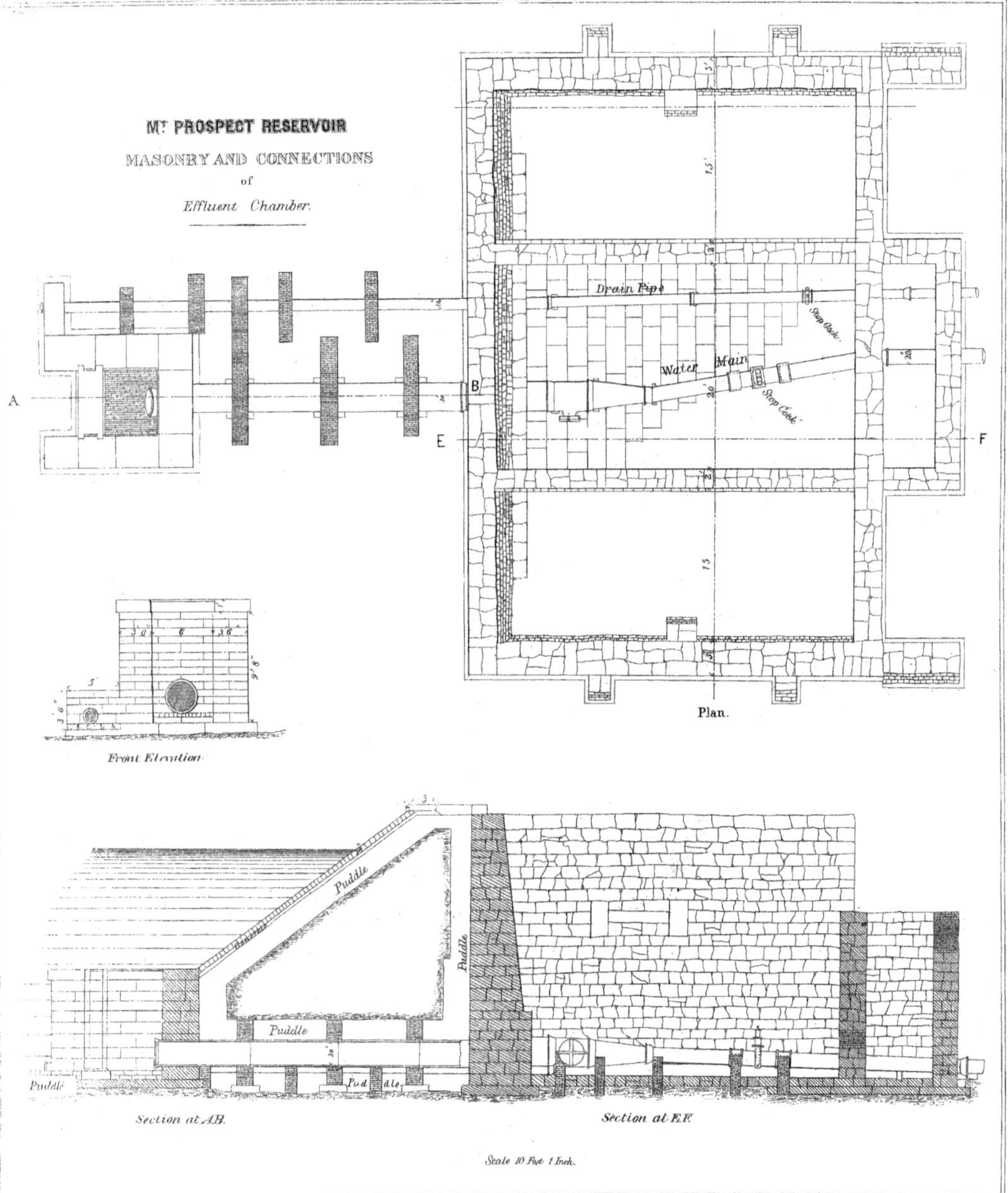


D. Van Nostrand, Publisher, 192 Broadway, N.Y.
J. Bien, Lith. N.Y.

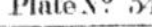


# SECTION OF GATE HOUSE
at
## Mt. PROSPECT.

*Scale 8 Feet = 1 Inch.*

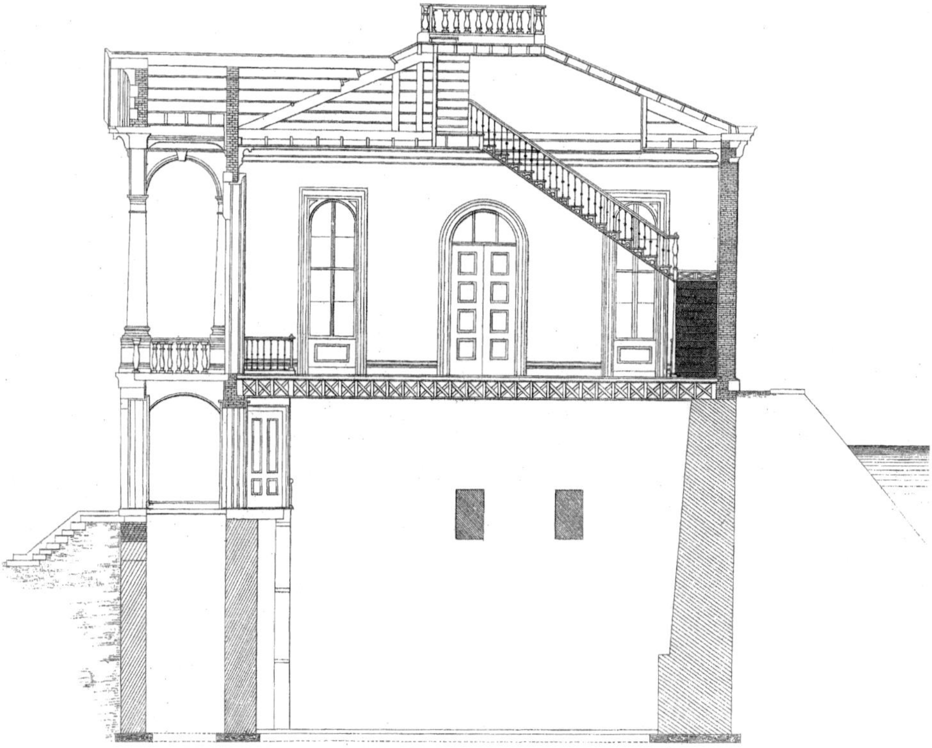

D. Van Nostrand, Publisher 192 Broadway N.Y.

J. Bien Lith. N.Y.

# Sections of Egg Shaped Sewers.

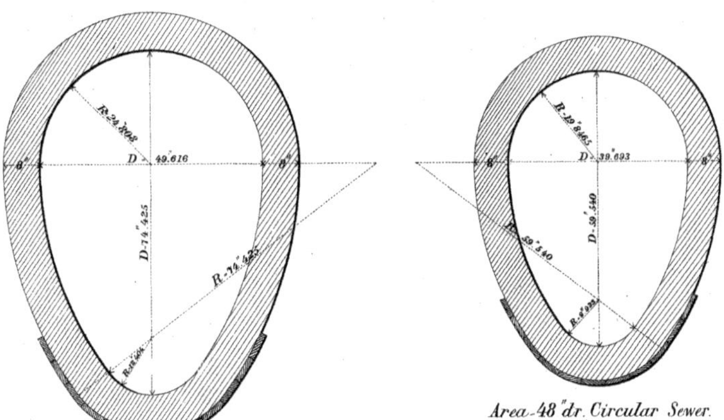


Area-60" dr. Circular Sewer.

Area-48" dr. Circular Sewer.

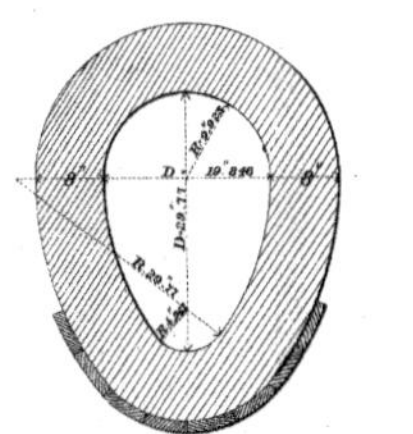

Area - 24" dr. Circular Sewer.

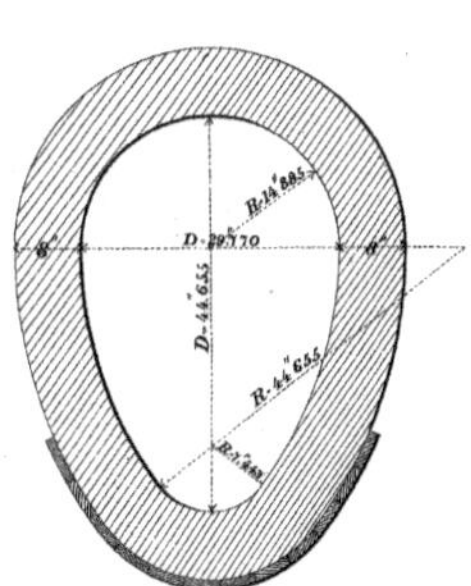


Area - 36" dr. Circular Sewer.

Scale 2 ft. = 1 Inch.

D. Van Nostrand, Publisher 192 Broadway N.Y.

J. Bien Lith. N.Y.

# MAN HOLE
*and*
## STREET BASIN.

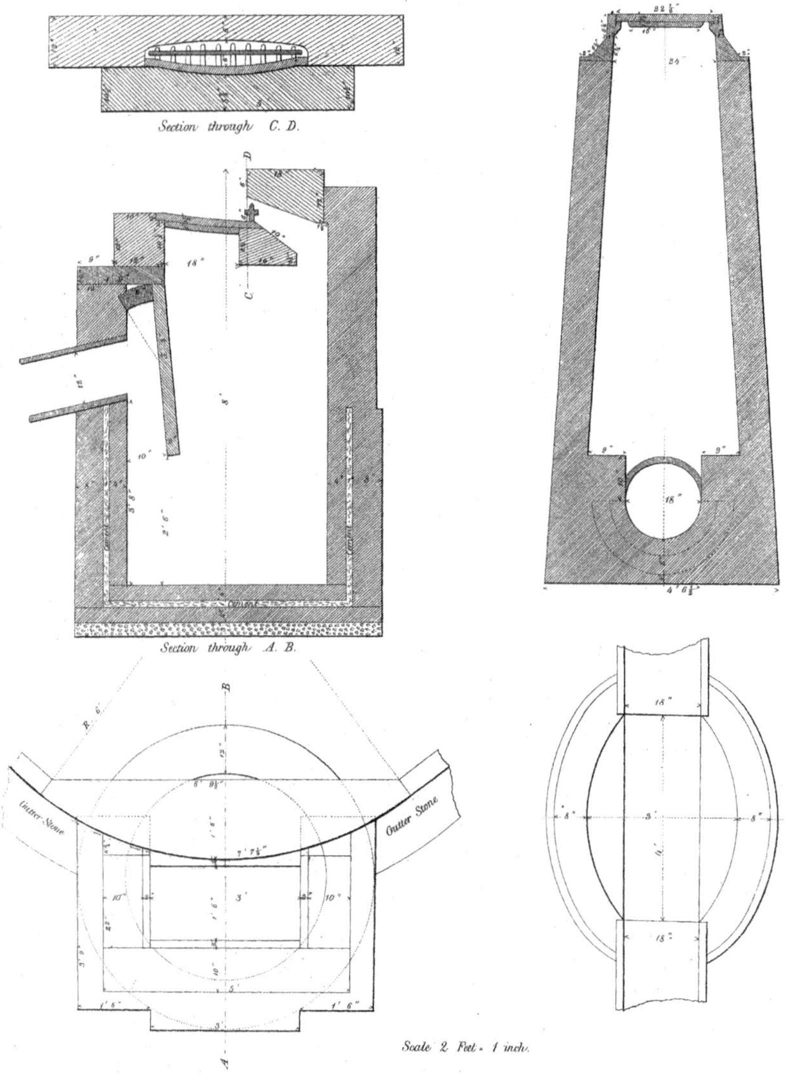

*Section through C. D.*

*Section through A. B.*

*Scale 2 Feet = 1 inch.*

D. Van Nostrand Publisher 192 Broadway N.Y.

J. Bien Lith. N.Y.

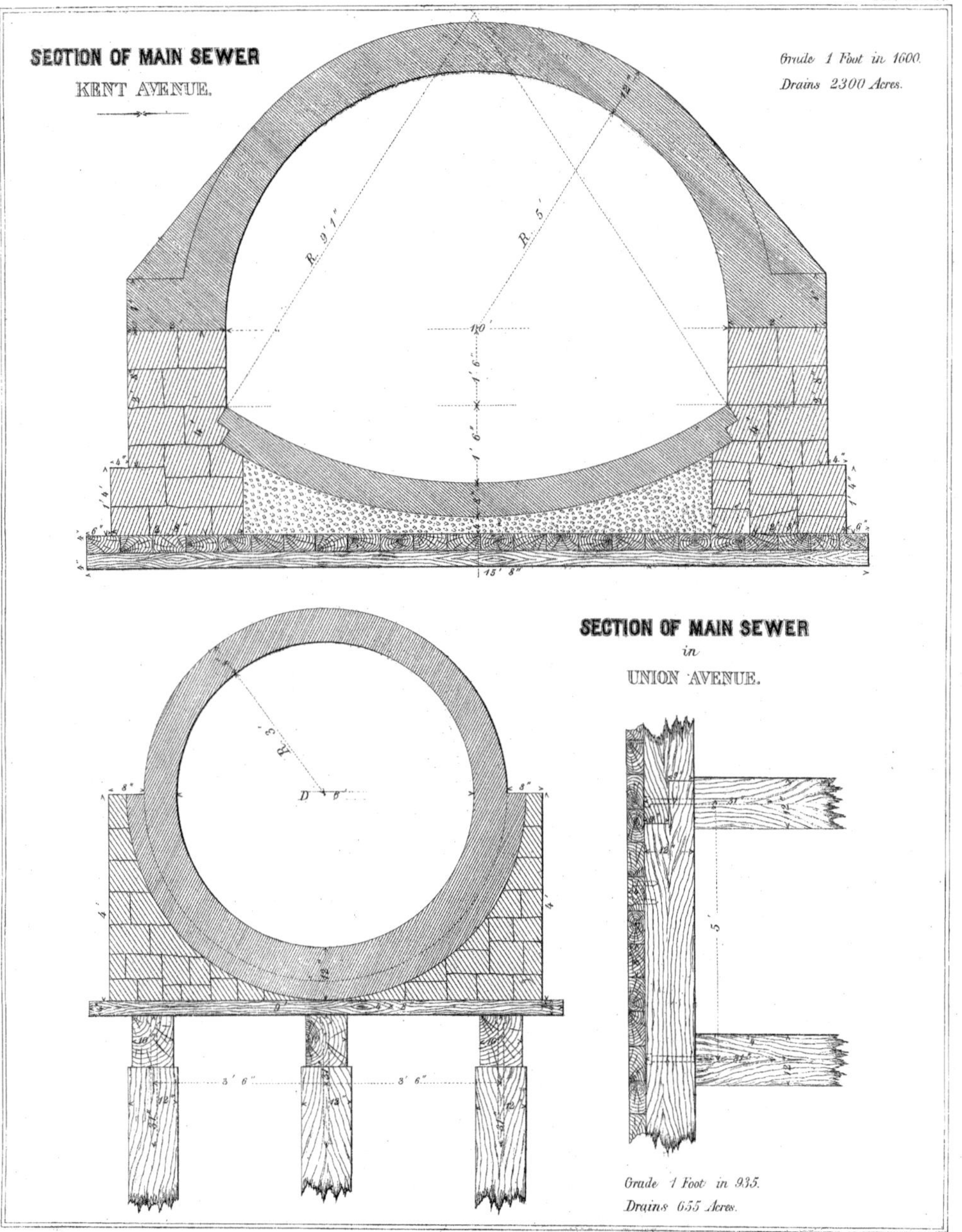


D. Van Nostrand, Publisher 192 Broadway, N.Y.

J. Bien Lith. N.Y.

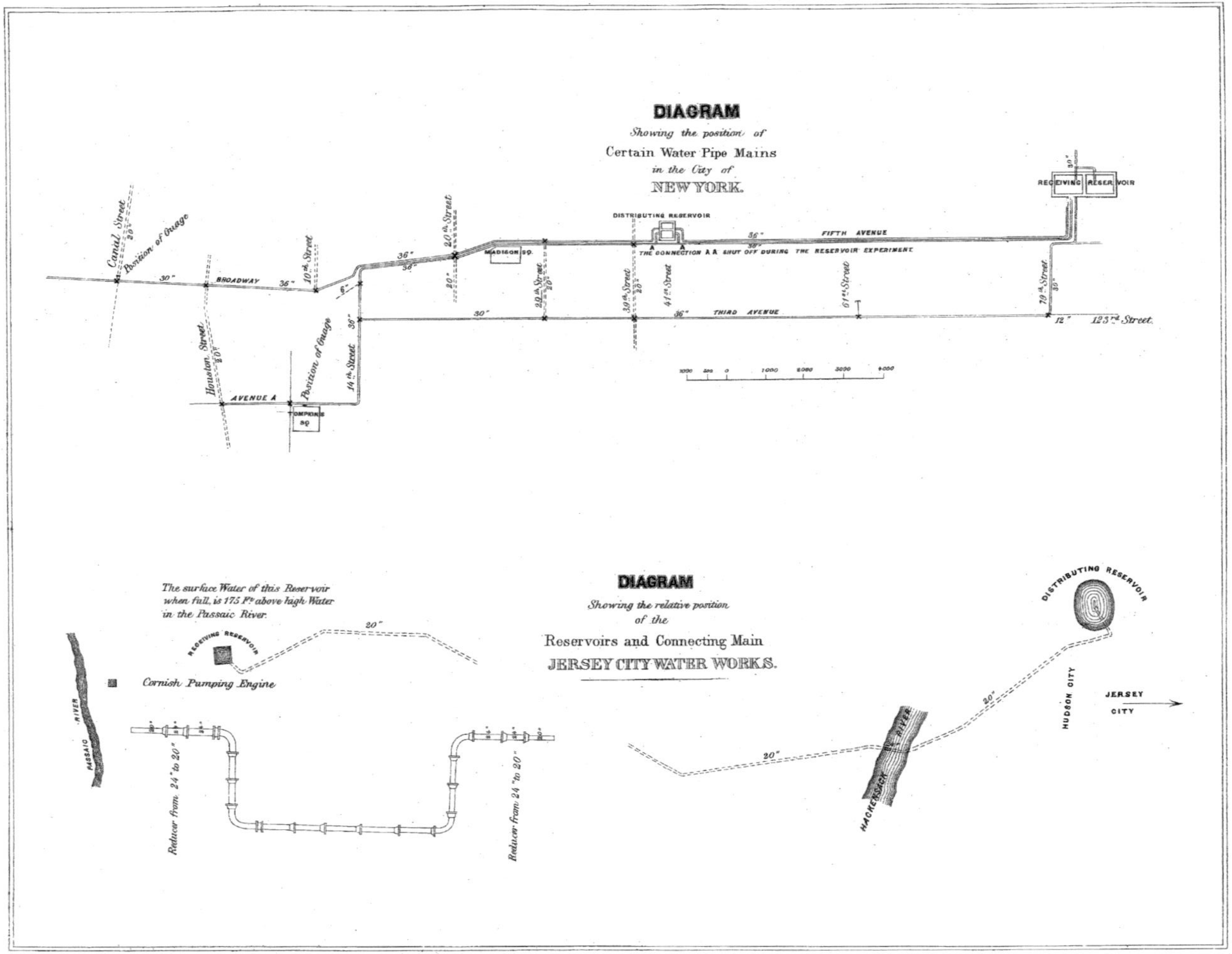

## DIAGRAM

*Showing the position of*
Certain Water Pipe Mains
*in the City of*
NEW YORK.

DISTRIBUTING RESERVOIR
RECEIVING RESERVOIR
FIFTH AVENUE
THIRD AVENUE
BROADWAY
AVENUE A
THE CONNECTION A A SHUT OFF DURING THE RESERVOIR EXPERIMENT.
Canal Street
Position of Guage
Houston Street
10th Street
14th Street
20th Street
29th Street
39th Street
41st Street
61st Street
79th Street
123rd Street
MADISON SQ.
TOMPKINS SQ.

## DIAGRAM

*Showing the relative position*
*of the*
Reservoirs and Connecting Main
JERSEY CITY WATER WORKS.

*The surface Water of this Reservoir when full, is 175 Ft. above high Water in the Passaic River.*

RECEIVING RESERVOIR
*Cornish Pumping Engine*
PASSAIC RIVER
*Reducer from 24" to 20"*
*Reducer from 24" to 20"*
HACKENSACK RIVER
DISTRIBUTING RESERVOIR
HUDSON CITY
JERSEY CITY

D. Van Nostrand, Publisher 192 Broadway, N.Y.
J. Bien Lith. N.Y.

www.ingramcontent.com/pod-product-compliance
Lightning Source LLC
LaVergne TN
LVHW010221110826
845151LV00004B/1165

* 9 7 8 1 4 2 5 5 2 7 7 7 8 *